The Science behind Global Warming

and Fossil Fuels Replacement

L. M. Rose

email: chemeng@btinternet.com

© Chemeng, Beaminster, UK

Dr L. M. Rose obtained a BSc degree in Chemistry and a PhD in Chemical Engineering, researching mass transfer, from Birmingham University, UK.

He spent 15 years with large companies in the chemical and oil industries, working on engineering design and process development. Following this was 15 years in the ETH, Zurich, the Swiss Technical University, in the Systems Engineering group, teaching and researching in mathematical modeling, statistics and business studies.

Since then he has been working as a consultant in the area of modeling of chemical systems and process development, as well as managing his company markeing engineering software.

His wide range of experience makes him eminently suitable to understand and comment on the science behind global warming as well as to judge the measures being considered to combat it.

Books by the same author –

The Application of Mathematical Modelling to Process Development and Design – Applied Science Publishers, 1974

Engineering Investment Decisions – planning under uncertainty, Elsevier, 1976

Chemical Reactor Design in Practice, Elsevier 1981

Distillation Design in Practice, Elsevier, 1985

The Art of Process Design (co–author G L Wells) Elsevier, 1986

Preface

The problem of global warming and the change in energy sources is one of the major problems facing the world today. Despite its importance, the gravity of the problem is under dispute.

Meteorologists warn about unacceptably high temperatures. biologists warn about threatened ecosystems on land and in the sea, oceanographers warn about changes in ocean currents, and the consensus that there is a problem is confirmed by the general report that '*the majority of eminent scientists*' believe that it is serious. However there are others, meteorologists amongst them, that are not so sure.

There is a element of dispute between the scientists – this is bad for science. Scientists should agree where the evidence is clear, and identify areas of uncertainty for further study. There should be no room for scope for 'disagreements'.

I have often be asked what I thought about global warming, and since I had no view, I though I should try to understand the problem. So began a 2 year study to try to resolve this magnitude of the problem, from a scientific basis, going back to the original scientific papers wherever possible. This text is the result of this study.

It took me from mass balance studies (chemical engineering), radiation (spectroscopy), sea water carbonate equilibria (chemistry), through to global modeling (climatology, mathematics) past climates (palaeontology), to energy sources (engineering) and their implementation (economics) before I felt I properly understood the problem.

It did occur to me that there must be very few people who have studied the whole range of subject and who are therefore in a position to make reasoned judgment. Least likely to get it right are *eminent* scientists, who have probably excelled in only one subject, and that not a relevant one.

My journey has been quite fascinating. Our CO_2 emissions are causing concentration increases into the unexplored levels; the sea is absorbing more than one might expect; the CO_2 itself does not produce very much global warming and, most probably, fossil fuels will become scarce before the planet suffers ill–effects from higher CO_2 levels. We must move away from fossil fuels, because they will become too expensive or be no longer available. Renewable supplies are thwart with problems when they are more that 10 –20 percent of our total energy supply. What is desperately needed is a replacement large scale, continuously operating source for generating electric power.

Nuclear power is the most likely solution, and by moving away from uranium based nuclear power, and by incorporating recycling of waste fuel it should be possible to provide safer, cleaner, cheaper, sustainable nuclear power.

There needs to be more development and very careful decisions made on our future energy supply. There is no need for haste, It is more important to get it right than to meet arbitrary targets.

All these conclusions backed by careful argued scientific evidence, with quantitative numerical analysis. If you want to see the arguments then please read on.

Acknowledgments

The availability of Wikipedia has been much appreciated. This marvelous concept enables new subject areas to be entered painlessly, because its short description are a good first start to any new theme being studied. Thanks to Wikipedia also for most of the illustrations in Chapter 9. Likewise the Google Scholar is an amazing facility, enabling the user to locate scientific papers without having to visit National Libraries. Google Scholar finds relevant papers, provides links to abstracts where, in some cases, the whole paper is available to be read.

The Amazon company and its facility for self–publishing, *Createspace,* is also much appreciated. The traditional procedure of having to find a publisher who thinks he might make a profit out of the text before it can be published, does not sit well with the author who wrote it because he simply feels it is something that needs to be said. The procedure with Amazon, of being able to distribute the text without high initial costs is to be applauded. It may well have repercussion in the whole publishing industry.

I would like to thank Professor John Barnard for his reading of the draft manuscript and his useful comments thereon. Many thanks also to my wife, Jane Rose, for undergoing the tedious task of proof reading.

Murray Rose
March 2013, Beaminster, UK

Table of Contents

Chapter 1   Our Planet, CO₂ and Fossil Fuels

Global warming is one of the most discussed problems of the 21st century. Ever since the realization of the relentless increase of CO_2, as shown by the Keeling curve, there has been concern with the effect on our world. As CO_2 is known to upset the radiation balance of the planet by absorbing outgoing radiation, there is the worry that the planet will increase in temperature, and its delicate biosystem will not be able to cope with the changes that occur.

This matter has been taken up by environmentalists, and various pressure groups have made it quite clear that they believe we are heading for global disaster. The environmentalists include botanists and zoologists, who are only too willing to warn of the dangers of climate change and the effect on species. Warnings are dire: changes will occur much faster than evolutionary developments can follow; mass extinctions can be expected.

The problem has been taken up by our politicians, whose duty it is to look after our future and this has resulted in a host of international conferences and decisions to limit our CO_2 emissions, albeit, without mentioning that we should limit growth itself.

Committees have been formed to guide the politicians. The most notable being the IPCC (the International Panel on Climate Change) which has been formed to warn of possible future climate problems due to fossil fuel burning. National committees have been formed by politicians to advise on how to implement the policies decided by the international committees. All these committees have very specific remits. The IPCC remit is to warn of possible future dangers, and this they do with enthusiasm in all their reports   Committees such

1

as the CCC (Committee for Climate change, UK) have the remit to decide on how to meet the politically agreed CO_2 targets, with little concern for the science. Climatologists and meteorologists are interested in ensuring possible future difficulties are understood. There is no group whose remit it is to predict the most likely future as realistically as possible.

Politicians have to show that they are men of action, able to tackle important problems and solve them. To show leadership and action they must make coordinated decisions at world conferences and return home to instigate agreed policies, with considerable vigor. Meanwhile, the economists, even the politician/economists, do show considerable doubts on renewable energy investments which are being made on borrowed money and which are guaranteed to make a loss, even though this activity does provide employment for millions. Engineers are also very doubtful of the wisdom of installing new forms of energy, as they are so much less economic and provide more problems than existing forms. Engineers are used to determining the best, most economic way of working, and then proceeding in that direction. The idea of choosing a poor substitute and then investing heavily in it is difficult for them to accept. So we see chinks of opposition to these policies; the world divides into those for and against global warming alleviation policies.

Over all this is the uncertainty of the science behind the future forecasts. Most people, the public and scientists, encouraged by the climatologists and meteorologists, accept the warnings without question. Other scientists, who also include some climatologists and meteorologists, do not believe that the matter is adequately understood, and are doubtful about the seriousness of the outcomes. So the world is divided into believers and deniers, even in the scientific community. This is bad for science, we should be able to have a unified, well–founded opinion, expressing real doubts where they exist. The aim of this book is to understand the problem, look at all the evidence and try to come to an independent, scientific conclusion.

So let's get started!

1.1) Carbon and Carbon Dioxide

Carbon is one of the most important elements on our planet and the one which makes our planet what it is. It is the basis of all living matter, and it one of the main constituents of the Earth's surface. It is used in providing us with the majority of our energy.

Let us start by looking at the element itself, and then consider the role carbon dioxide plays in our world.

Carbon is a very small element containing only 6 electrons. Its nucleus therefore needs 6 protons and usually has 6 neutrons, making carbon 12 (^{12}C), or sometimes there are 7 neutrons giving ^{13}C , or 8 neutrons which gives ^{14}C. All these isotopes – ^{12}C, ^{13}C, and ^{14}C – are important in our story. The 6 electrons mean that the second orbital contains only 4 electrons where it really needs 8 to be stable. This means there is ample opportunity to link with other atoms to fill these missing spaces. It is this lack of electrons which gives the opportunity for there to be millions of combinations with itself or with other elements, which result in carbon being involved in such a variety of substances such as diamond, limestone, rubber, proteins, sugars and the whole range of organic chemistry and living things.

Oxygen is also abundant on the planet, and forms a very stable compound with carbon, carbon dioxide. This is a combination of one atom of carbon with two atoms of oxygen, with two electrons from each oxygen atom combining with the 4 electrons of the carbon to make a molecular orbital containing 8 electrons – the preferred stable configuration.

Carbon dioxide (CO_2) reacts with water to form carbonic acid, and this acid reacts with alkali to form carbonates ($CO_3"$) and bicarbonates (HCO_3'). On our planet we have plenty of alkalies from calcium and magnesium and an abundance of water, which means that over the life of the planet the carbon finds itself combined as carbonates, mainly calcium carbonate ($CaCO_3$), in forms such as limestone, chalk and marble.

The history of carbon over the life of the planet is fascinating.

Chapter 1   Our Planet, CO2 and Fossil Fuels

After our planet formed, as it cooled, the water and carbon dioxide would have been the major components in the atmosphere. Oxygen is unlikely to have been present; as it is very reactive it would not be present as an uncombined element. It is generally agreed that some particular biological life form in the sea was instrumental in causing oxygen to enter the atmosphere, by breaking down CO_2 with some form of photosynthesis reaction, liberating oxygen. This gave the atmosphere its oxygen which is now the basis for most life forms. Over millions of years these life forms in the sea manufactured skeletons and shells from the carbon dioxide in the seawater. On the death of these organisms, the skeletal remains fall to the bottom of the sea and produce a sediment of calcium carbonate, and the organic matter from organisms fall to the sea bottom and is the source of the geological deposits of our crude mineral oil. As the result of continental drifts and upheavals of the earth's crust we now have these sedimentary rocks being found where they are – some on the sea bed, often on land and sometimes high in the mountains. The oil collects below sea level, between impervious strata of rocks.

This theory puts life forms at the center of the development of the planet. The planet is what it is because of the importance of the biosystems that have been present for billions of years. Life is an integral part of the planet.

This idea that our planet is a 'living system', with the biological systems being an integral part, which reacts to events, and compensates for changes, was proposed by James Lovelock. He named it the 'Gaia Hypothesis'. Lovelock is an eminent scientist, with significant developments in chromatography and the discovery of CFC's in the atmosphere to his credit. He has detailed knowledge of the theories of the beginning of the planet, and his ideas on the control systems operating are convincing;– he states: *a body that has existed for 4.5 billion years must have some inherent stabilizing mechanism – eg Gaia* . However right he might appear, his arguments have moved away from being scientific and have a mystic ring about them. For this reason he is not taken seriously by the scientific community, and is best not cited in serious scientific debate. He is best looked upon as a prophet – he may well be proved right, but should not be used in scientific argument.

There is still the natural cycle of carbon dioxide going on to this day. Carbon dioxide enters the atmosphere from volcanic activity, as carbonates on the sea floor are subducted into hot magma. The heat decomposes the calcium carbonate into calcium oxide and carbon dioxide, and this CO_2 is released, often with explosive violence.

Carbon dioxide is washed from the atmosphere by rain which eventually returns to the sea, after having flowed over limestone, picking up calcium in passing, in the form of calcium bicarbonate $(Ca(HCO_3)_2)$. This bicarbonate is needed by most sea life forms which then die and leave their skeletons 'raining' onto the sea bottom.

$$CaCO_3 \quad + \quad CO_2 \quad + \quad H_2O \quad \rightarrow \quad Ca(HCO_3)_2$$

calcium carbonate + carbon dioxide + water → Calcium bicarbonate
solid rock   *dissolves in rain*   *carried to the sea*

This reaction is known to us all as it is a common source of many domestic problems. The calcium bicarbonate is in the domestic water supply if the water has flowed over limestone before collection. It is this which makes hard water hard. It interferes with soap and precipitates as a scum of calcium stearate making washing difficult.

It is decomposed by heat to solid calcium carbonate

$$2\,Ca(HCO_3)_2 \rightarrow \quad Ca\,CO_3 + CO_2 + H_2O$$
dissolved   →   *solid scale*

and this produces lime scale on kettles, heating systems and boilers, and leaves a lime deposit on all utensils and surfaces.

The part played by CO_2 in plant life is twofold – that of being converted to sugars, and the reverse, that of being liberated by the oxidation of sugars to provide energy for the living organism.

The conversion of CO_2 and water into sugars by the addition of energy is very complex chemistry. The energy is sunlight, and the complex chemistry is photosynthesis, summarized as

$$6\ CO_2\quad + 6H_2O + energy \rightarrow \quad C_6H_{12}O_6 + 6O_2$$

carbon dioxide + water   + sunlight →   glucose   + oxygen

Animals consume plant sugars, dissolve them, and convert them into animal tissue, again using very complex chemistry. Or they can, together with oxygen, convert them back into CO_2 and water, with a release of energy, which is used to produce motion within their tissues. This reverse process also occurs in plants where it is called respiration.

On land as in the sea, plants use sunlight and photosynthesis and CO_2 in the atmosphere to produce plant material. As these plants die, they either decompose by biological activity back to CO_2, or are submerged and are finally converted by heat and pressure to fossil fuel – peat, lignite, brown coal, coal, oil, and natural gas.

These processes have maintained an equilibrium over millions of years with CO_2 formed geophysically being balanced by CO_2 removed biologically from the cycle by being buried as fossil fuels and carbonates. The amount of CO_2 in the atmosphere to maintain this equilibrium is tiny, with concentrations varying between 0.02 and 0.04 % .

1.2) Fossil Fuels

Into this delicately balanced, well–established system, comes mankind with a desire to have more energy. In prehistoric times this was achieved from burning plant materials, such as wood, which was recycling the CO_2 which would have been converted into CO_2 anyway by bacteriological decomposition. The established cycle was not upset. After 1700 water power and wind power were 'harnessed', taking kinetic energy from the water and air flows. Again the energy would dissipate itself anyway, and did not involve CO_2. This did not upset the established cycle.

Not until the use of coal as an energy source, around 1750, did the quest for energy interact with the CO_2 cycle. The burning of fossil fuels is effectively recovering the energy of the sunlight, stored in the fossil fuel many millions of years ago by reconverting the fuel back to CO_2. This CO_2 is released into the atmosphere and is a new input stream into our established system. As long as the quantities involved are small compared to the planet this is of little concern, but as the burning of fossil fuels increases it deserves more consideration. At what point do we need to be concerned?

The Appendix to this chapter gives a rough calculation of the magnitude of the CO_2 emissions, just using common experience, which shows convincingly that we are adding measurable amounts of CO_2 to our atmosphere, and so it should be a matter of concern.

The atmospheric CO_2 concentration used to be stable at 280 parts per million by volume (ppmv). The anthropogenic CO_2 is adding to this CO_2 at the rate of about 1% per year – set to double in 100 years time.

That such a simple calculation can show our present production of CO_2 is significant should convince any 'denier' that there is cause for concern and that our burning of fossil fuels is a significant problem. We will be showing later how more accurate calculations find levels of CO_2 in the atmosphere which agree exactly with the quantities of CO_2 we are releasing.

Central to the problem is the very low standing concentration of CO_2 in the atmosphere. This means that even if the amount of the CO_2 being added to the atmosphere is relatively small in planet dimensions, it is significant enough to alter the low level in the atmosphere.

From 1820s, coal was used in quantity as the industrial revolution took hold. Coal remained the main source of energy until the 1950s when oil took over, because of convenience and price. In 1990, the discoveries of large quantities of natural gas caused gas to become a major energy source, again on convenience and economic grounds. In all cases, our major energy sources are fossil based, which upsets the CO_2 balance on the planet.

There is also hydro and nuclear power, which do not involve emissions of CO_2. These will be discussed in detail in later chapters when we turn to alternatives available to us.

The different forms of fossil fuel from coal to gas do involve different quantities of CO_2 being released per unit of energy produced. This is merely because of their different chemical composition. Coal is mainly carbon, so the majority of the energy comes from carbon conversion to CO_2. Oil is hydrocarbon, so some energy comes from converting the carbon to CO_2, but some from the conversion of the hydrogen to water. Gas is again a hydrocarbon, but with a greater hydrogen to carbon ratio, so that more of the energy comes from the production of water, without association with CO_2.

Table 1.1 compares the different CO_2 emissions for equal energy release from the various fuels

Table 1.1
Comparison of Fossil Fuels
Heat of Combustion and CO2 Emissions

Fuel	Heat of Combustion kWh/kg fuel	CO2 Emission kg CO2 /kWh
Natural gas	13.9	0.198
Gasoline	12.5	0.247
Fuel oil	12.6	0.25
Coal	10.2	0.336

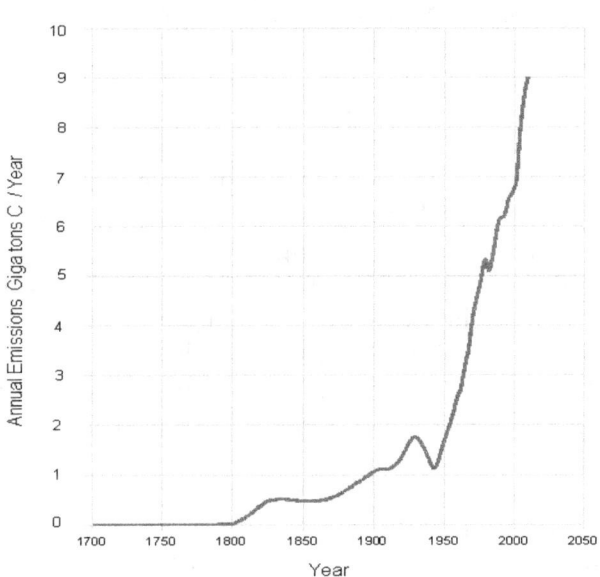

Figure 1.1 *Shows the growth of carbon released into the atmosphere by the burning of Fossil Fuels, from 1700 to the present day. Data taken from USA source (CDIAC).*

So it is true that by switching from other fossil fuels to gas, the carbon emissions will be reduced, and CO_2 emission targets can more easily be achieved. As the gas reserves are used up, there must be a switch to oil, and then to coal, so in the long term picture, little will have been achieved, unless moves are made to look for energy supplies elsewhere, which are not fossil fuels.

To be more scientific we should return to those organizations that have dedicated themselves to providing accurate well justified data. The Carbon Dioxide Information Analysis Center (CDIAC) is an institution in the US which provides the history of global fossil fuel usage. They produce an updated estimate of annual CO_2 releases from fuels for the whole planet. Figure 1.1 is from their data. It is the best data available to show the trends in CO_2 from fossil fuel usage by mankind on the planet for his energy needs.

1.3) The Growth of Atmospheric CO2

A second institution of great important in this study is the **Mauna Loa Observatory, Hawaii, U.S.A.**

Associated with the Scripps Institute of Oceanography. In 1958 C. D. Keeling first measured the atmospheric CO2 levels on a routine basis and found the value increasing with time (Keeling, 2002). The 'Keeling' curve is basic to all CO2 and global warming studies. Figure 1.2 shows a record of the individual measurements, taken from 1958 to 2004. The annual cycling is thought to be due to the interaction between the CO2 in the atmosphere and growing season of plants. The steady growth is the subject matter of this book.

We can compare the data from the Keeling curve with our rough calculation of expected anthropogenic CO2 growth of 2.2 ppmv per year. Consider the years 1980 to 2000, the Keeling curve shows the CO2 composition of the atmosphere changing from 334 to 367ppmv, equivalent to a rise of 1.6 ppmv per year. Surely this is near enough agreement to convince even the most skeptic that our CO2 emissions are causing the rise shown by the Keeling curve.

If the anthropogenic (man—made) CO2 of Figure 1.1 is the explanation for the CO2 rise in Figure 1.2, then the two should be closely related. This can be shown by considering one point – say 1990:

> The carbon global emissions were 6.0 billion tons (6.0 giga tons) carbon per year – this is equivalent to 6 x 44/12 = 22 billion tons CO2. As we have shown , the Keeling curve shows the rate of increase around 1990 was 1.6 ppmv. Scaling from our earlier calculation this is equivalent to an amount
> 22 x 1.6/2.2 = 16 billion tons CO2.

The figures we have used here are best—estimate published figures, showing only 16 out of 22 billion tons of CO2 (72%) of the released CO2 remain in the atmosphere. This must be because the CO2 has found other places to go – so is it being absorbed by the oceans? is some being captured on land by the soil and vegetation?

These matters will be returned to in later chapters.

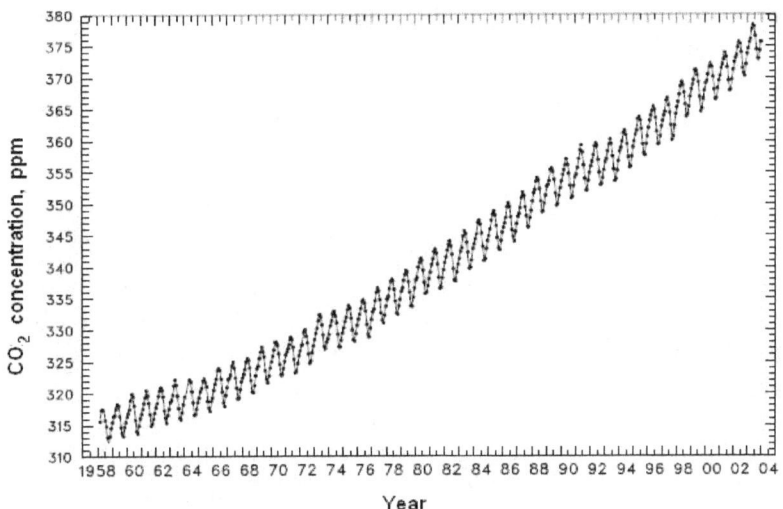

Figure 1.2 The Keeling Curve. Growth of CO2 in the atmosphere
Observed data taken from measurements at Mauna Loa (Scripps), 1958 – 2004.
It was this work which alerted scientists to the growth of anthropogenic CO2.

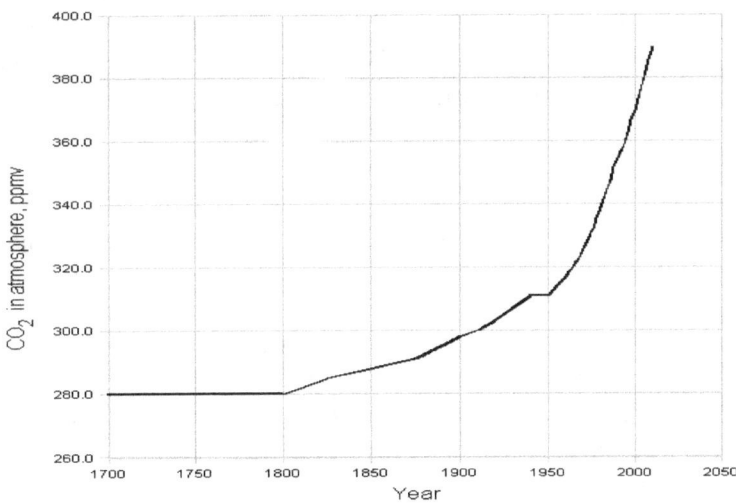

Figure 1.3 The growth of CO2 in the atmosphere since the industrial revolution.
Before 1958 the data has come from air bubbles trapped in ice. After 1958 the data is measured directly, and comes from figure 1.2.

Figure 1.3 is the best estimates of the whole change in atmospheric CO_2 since the onset of the use of fossil fuels, of which the Keeling curve is a small, but accurate segment

1.4)   Carbon Isotope Measurements

Isotopes will be described in detail in chapter 8, but are mentioned here because they provide evidence of the source of CO_2. The normal carbon ubiquitous on the planet has a small part as isotope ^{13}C, slightly heavier than the normal isotope ^{12}C. This weight difference causes it to react at slightly slower rate than the normal ^{12}C, so there can be a depletion in the ^{13}C in the reaction product, compared to the original material. In the case of photosynthesis, plants absorb ^{13}C slightly slower than the isotope ^{12}C, so plant material contains less ^{13}C than the normal level. This leads through to the fossil fuels, which also contain less than the normal levels of ^{13}C. Burning this fuel produces CO_2, with less than the normal ^{13}C levels. (Zeebe, 2001)

As the fossil fuel CO_2 mixes with CO_2 from the planet's atmosphere, one would expect the ^{13}C content of atmospheric CO_2 to steadily fall. Figure 1.4 shows that this is indeed the case.   The atmosphere used to have a ^{13}C content 2.41% below the 'reference carbon ' . Plant carbon contains ^{13}C 4.4% below the ' reference carbon ', so as more plant CO_2 enters the atmosphere, the ^{13}C should fall from −2.41 towards −4.4%;

In isotope chemistry the conventions are defined as changes from a reference material,(δ), and the changes are quoted in parts per mil(‰), not percent (%). So the atmospheric δ value of −24.1 ‰ ^{13}C mixed with plant carbon contains a δ value of −44 ‰ ^{13}C.

Looked at quantitatively, figure 1.3 shows the CO_2 has risen from 280 ppmv in 1850 to 367 ppmv in 2000, there has been a 30% increase in CO_2. If this comes from fossil fuel, with a 20 ‰ difference in ^{13}C, then the mean atmospheric isotope ^{13}C will be increased from −24.1 to −29 ‰. The actual measurements given in figure 1.4 show the ^{13}C has risen from −24.3 in 1900 to −28.8 ‰ in 2000 − surprisingly good agreement! (Andres, 2011)

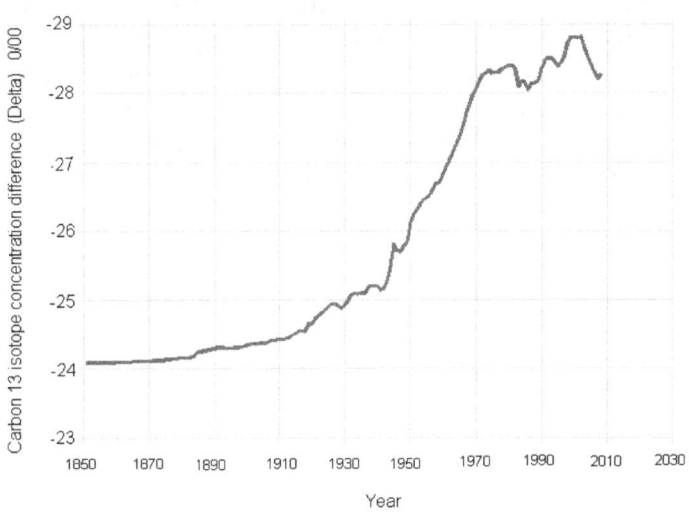

Figure 1.4 **The change in isotope ^{13}C on Atmospheric CO2 over the last 150 years.**
The CO2 from fossil fuels contains less ^{13}C than early atmospheric CO2. Over time this atmospheric ^{13}C has increased and the change is remarkably similar to the emissions of Anthropogenic CO2 curve – Fig 1.1 . This is considered strong evidence that the increasing CO2 is from fossil fuels.

1.5) Conclusions

The aim of this first chapter has been to show the convincing arguments that the rise in CO2 has been due to the use of fossil fuels and is man–made. The shape of the curves show CO2 growth and fossil fuels use are very similar. Looked at quantitatively, the numbers are in good agreement. The 'fingerprint' evidence left by isotope ^{13}C is very convincing, and the numbers again are in good agreement. This does mean that we are on track for doubling the CO2 levels in less than 100 years. We are moving into unknown territory, producing levels of CO2 which have not been on the planet for some millions of years.

There are no competing explanations for the CO2 increase, and so it is hoped that all readers can accept this evidence; there should be no deniers at this stage.   This only relates the CO_2 growth to anthropological activity. It does not provide any evidence that this growth is detrimental, or is the cause of global warming. CO_2 increases may even be advantageous, Later chapters will be discussing what effect this CO_2 rise is likely to have on the climate.

Appendix to Chapter 1

A Rough Estimation of the Magnitude of the Anthropogenic CO_2

We could simply rely on the experts to tell us how significant is our emission of CO_2, or we can roughly estimate for ourselves how important is our input of CO_2 . It is always more convincing when we understand where the figures have come from. So let us try to estimate from day to day experience whether our CO_2 input is a significant quantity in terms of the amount already in the atmosphere. Some of these figures might just be rough guesses, but the aim is to show how easy it is to get a feel for the magnitudes involved. As a control, the figures are compared with published accurate figures throughout the calculation.

Let us consider a European couple with a car driving 12,000 miles, per year, with their house being heated with as much oil as the car uses, and as much again is spent on other energy usage. Assume also that the total energy used by the whole country is twice the personal usage to include energy used for industry, transport, and all other non–domestic usage.

The couple would be using about:
12000/45 x 4.5 =   1200litres = 1000 kg/yr gasoline in the car
(at 45 miles/gallon or 6 litres/100 km)

Total personal use by the couple for all energy being 3 times this:– ie 3.0 tons/yr gasoline equivalent.

One mole of gasoline C_8H_{10} (molecular weight 106) burns to produce 8 moles of CO_2 (molecular weight 44)
so, 3 tons of gasoline will produce

$$3 \text{ x } 8 \text{ x } (44/106) = 10 \text{ tons of } CO_2 \text{ per couple}$$

now we assume the state will use an equal amount, so our couple will be causing 20 tons to be emitted – or 10 tons per person.

(A figure for Europe of 10.7 tons per year per person has been published)

The UK has a population of 55 million, so the total emissions for the UK according to this calculation are 55 x 10= 550 million tons

(published estimates are 700 million tons)

For the whole world with a population of 7 billion, we can estimate the total anthropogenic CO_2 emitted if we can assume a suitable average emission per person . Let us assume the world average is 30% of the European figure – 3 tn CO_2/year per person

This would give an annual global emission of CO_2 of 21 billion tons

(published data reports 22 billion tons CO_2 /year)

Now if this were all to go into the atmosphere, what concentration change would it give?

Firstly, what is the weight of air on the planet?

The diameter of the globe is 12,800 km. So, remembering the formula $a = 4\pi r^2$, the surface area is:

$$3.148 \times 12,800 \times 12,800 = 5.14 \ 10^8 \text{ sq km}$$

Now the weight of air per square inch 14.7 lbs (as all steam engine buffs know, this, the old definition of atmospheric pressure is the weight of air causing the pressure)

So the total weight of air is

$$(\text{km}^2 \text{ area x m/km x air pressure x in}^2/\text{m}^2 \text{ / lbs/ton})$$

$$5.14 \times 10^8 \times 10^6 \times 14.7 \times 1550/2204 = 5.31 \times 10^{15} \text{ tns}$$

(published data gives 5.13x10¹⁵ tns)

Distributing this CO2 into this mass of air gives a concentration change of

$$21 \times 10^9/5.31 \times 10^{15} \times 10^6 = 3.5 \text{ ppm by wt}$$

and converting this into volume measurements by correcting for molecular weights

$$3.5 \times 28/44 = \textbf{2.2 ppm} \text{ by volume (ppmv)}$$

The atmospheric CO2 concentration used to be stable at 280 ppmv, which suggests, even by this very simple calculation, the anthropogenic CO2 is adding to the atmosphere at the rate of nearly 1% per year – ie set to double in 100 years time.

(This 1% per year growth is often reported in the published data)

17

References

The Carbon Dioxide Information Analysis Center (CDIAC) http://cdiac.ornl.gov/

C.D. Keeling and T.P. Whorf ,Atmospheric CO_2 from Flask Air Samples at Mauna Loa Observatory, Hawaii, U.S.A. (2002), Scripps Institution of Oceanography, http://sio.ucsd.edu/

IPCC, 2000, *Emission Scenarios*– Nebojsa Nakicenovic and Rob Swart (Eds.), Cambridge University Press,

R.J. Andres, T.A. Boden, and G. Marland, *Annual Fossil–Fuel $CO2$ Emissions: Global Stable Carbon Isotopic Signature,* CDIAC, DOI: 10.3334/CDIAC/ffe.db1013. 2011

Scripps Institute of Oceanography, Mauna Loa Observatory, Hawaii, http://scripps.ucsd.edu/

Zeebe R.E., and Wolf–Gladrow D., (2001) CO_2 *in seawater:equilibrium, kinetics and isotopes*, Elsevier, Amsterdam

Further Reading

Gaia, *James Lovelock's The Ages of Gaia*, 1989, http://erg.ucd.ie/arupa/references/gaia.html

The Carbon Dioxide Information Analysis Center (CDIAC) http://cdiac.ornl.gov/

Scripps Institution of Oceanography University of California, San Diego, La Jolla, USA, http://sio.ucsd.edu/

IPCC, Working Group I, to the Fourth Assessment Report of the Intergovernmental Panel on Climate Change, *The Physical Science Basis,* 2007 Solomon, S., D. Qin, M. Manning, Z. Chen, M. Marquis, K.B. Averyt, M. Tignor and H.L. Miller (eds.)Cambridge University Press.

Chapter 2   Planet Temperature and Greenhouse Gases

2.1) Heat Transmission

Heat is a form of energy within matter, stored as kinetic energy in the molecule. The molecules are continually in movement and the atoms within the molecule are also 'vibrating'.   All this involves energy of motion and deformation above a minimum energy state. This energy indicates itself as the temperature of the material and the higher the temperature, the more movement and vibration is possessed by the molecule.

As the molecules are adjacent to each other, the energy is passed from one molecule to another by collisions, so that, in sufficient time, the whole body averages out to an even temperature. If there are permanent cold and hot positions in the body, then the energy moves from the hot to the cold by this interaction between adjacent molecules. This type of energy or heat transfer is called heat transfer by *conduction*. It can only occur when molecules can interact with each other.

If a solid body at one temperature is surrounded by a fluid, be it gas or liquid, at a lower temperature, than the conduction heats the surrounding fluid – this warmer fluid becomes lighter and flows upwards, being replaced by cold fluid. Hence heat is removed by the motion of the fluid and this is a second way of transferring heat – called transfer by *convection*.

If the body is in a vacuum, then heat transfer from that body by conduction and convection cannot occur, but there is a third method of heat transfer which still enables energy to be transferred. This is heat transfer by *radiation*. All matter emits electromagnetic radiation. This can be observed by feeling the heat radiated from a fire, or a hot cooker plate, or the sun. Radiation can be easily detected from hot bodies, but in fact there is radiation coming from all bodies, even cold ones, though this is more difficult to detect.

The type of electromagnetic radiation – be it white light, red light, or longer than red wavelengths (infra red, IR), depends upon the temperature of the material. A body emits white light when 'white hot' – for instance a light bulb filament; red when the body is 'red hot' – for instance an electric fire bar; and at lower temperatures radiation is still emitted but in the range of infra–red (IR), wavelength, which cannot be seen by the eye.

Besides the temperature affecting the 'color' of the radiation, the amount of radiation (intensity) also depends on the temperature of the material. A very hot body emits a high intensity of short wavelength radiation, a cold body emits a low intensity of long wave radiation

2.2) Infrared Radiation

So heat transfer by radiation is the way energy is transferred to a body despite it being in a vacuum. Heat transfer by radiation occurs all the time for all materials, but at ordinary temperatures it is much less significant than conduction and convection, and so is generally not important. However, it is important for our planet because it is the only way it can receive and remove energy, since it is space.

The classic starting point for studying radiative heat transfer is the 'black body'. This is a solid which absorbs all radiation falling on it, whatever its color (wavelength); it does not reflect any, but absorbs all.

Stefan (1879), and Boltzmann discovered that a black body emits a specific spectrum of radiation (a range of wavelengths or colors) – see figure 2.1 , with a characteristic shape. This figure shows how the the peak of the curve moves to higher emissions and lower

20

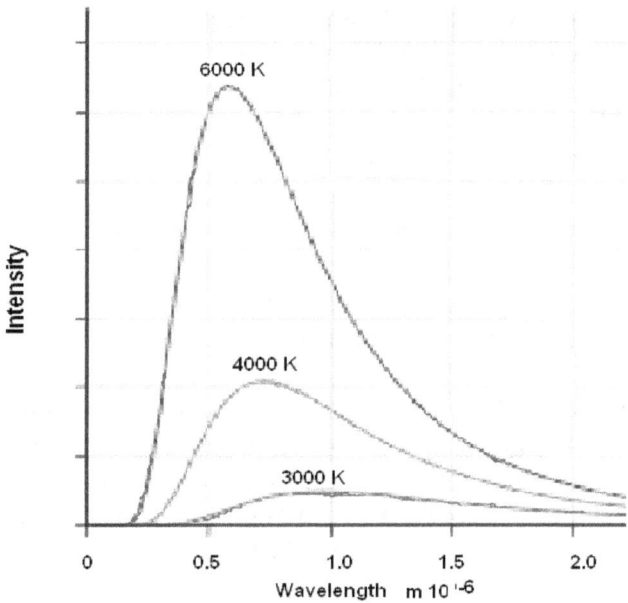

Figure 2.1 Black body emission curves at different temperatures.
All black bodies emit a characteristic spectrum which depends only on temperature of the black body. The 6000K line represents emission of white light from the Sun. As the body temperature drops, the intensity reduces and the peak moves to higher wavelengths. Around 300K the peak is a wavelength of 10 cm, and the intensity is 1/100,000 of the intensity of the Sun. These curves are given as background on figures 2.7 an 2.8 as they provide useful information on the temperature of the material emitting the radiation.

wavelength the hotter the body. This is a universal pattern, which holds true throughout the universe, from planetary behavior to the operation of domestic electric fires.

The sun is at a temperature of about 5500 °C, and the radiation it emits is white and intense. The Earth is at about 15 °C and the radiation it emits is low and in the IR range, not visible to the human eye.

There are mathematical functions available which describe the specific curves shown in Figure 2.1 and others which give total energy emitted as a function of temperature. It is found that the total energy emitted by black body radiation (I) is proportional to the temperature of that body raised to the fourth power:–

$$I = \sigma\, T^4 \quad W/m^2 \qquad\qquad (2.1)$$

where σ is the Stefan–Boltzmann constant with a value of
$$\sigma = 5.67 \times 10^{-8} \ \ W\,m^{-2}\,K^{-4}$$
This is a very important relationship which shows that the radiation emission is very sensitive to temperature level.

For those interested more in the theory, the appendix to this chapter shows that it is possible to calculate the mean temperature of the planet to balance the heat it receives from the sun. The resulting temperature is found to be –18 °C. But the planet mean temperature is nearer +15°C; there is an error of 33°C. We now have to explain this discrepancy.

2.3) The Greenhouse Effect

The reason for this discrepancy of 33° C is that the planet is covered by a layer of gas, – our atmosphere. If our atmosphere let all the IR radiation from black body Earth through into space, then the temperature would be 255K (–18°C) , as calculated. However, the atmosphere contains a number of gases, which absorb some IR. We need to look closely at the effect of this partial absorption of IR.

Let us now consider a planet surrounded by an atmosphere. To make it simple, assume that this atmosphere is a well mixed layer of gas. IR from the earth's surface passes through this layer, and a certain fraction (ε) is absorbed.

Everything, including gases, that absorb radiation, emits radiation and with emission coefficients (ε), the same as the absorption coefficients as long as the temperature is the same. A basis of all radiation theory is that the emission pattern corresponds exactly to the absorption patterns. A body at a constant temperature is receiving radiation from its surroundings, but is emitting exactly the same amount of radiation – otherwise it would not remain at a constant temperature.

But the difference between absorption and emission is that the absorbed radiation is in only one direction – the radiation going upwards, leaving the earth – but the emission is in two directions, up and down, – from both the top and the bottom of the gas layer. So the earth's surface emits radiation corresponding to the surface temperature (Ts) in an upward direction, which is absorbed by the layer of atmosphere which re–emits radiation both upwards and downwards corresponding to the temperature of the atmosphere (Ta). The surface temperature (Ts) is the result of the balance between all these radiation flows. If there are any changes in absorption, then Ts is 'forced' to change to maintain the balance. Figure 2.2 displays the radiation flows involved.

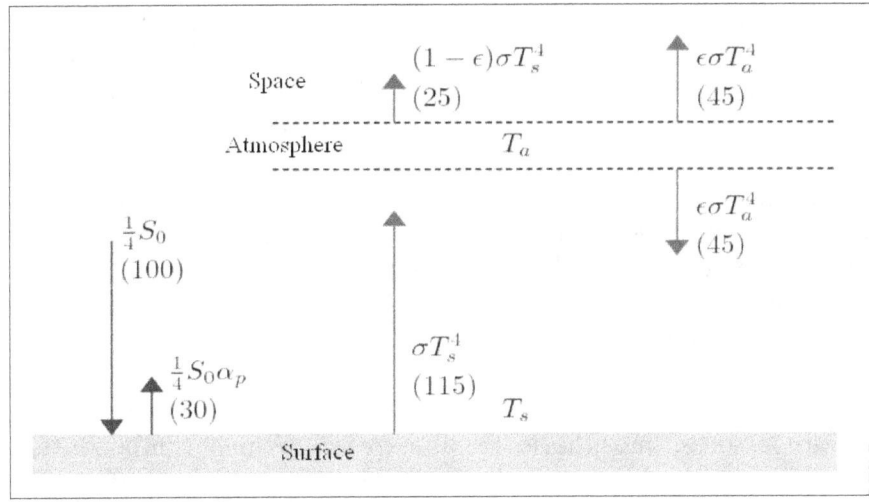

Figure 2.2 Energy streams to be balanced around the planet surrounded by its atmosphere.

The two left hand flows represent solar short waves, and the other four emitted infrared radiation. This simplified diagram represents the atmosphere as a single isothermal layer. The figures in brackets are intensity, expressed as % of incoming radiation

Again for those interested in the mathematics, Chapter 2 Appendix shows that the mean planet temperature can be calculated, and shows how much IR is absorbed and re–emitted by natural greenhouse gases (GHGs), and that the earth's temperature can be recalculated to show it is in fact 15°C

The analysis shows that our atmosphere absorbs 78% of the IR from the earth, and re–emits this in both directions. It is this re–emission in the downward direction which has given rise to the name of the phenomena as the 'greenhouse effect'. The planet has to thank the greenhouse effect for making the temperature high enough for it to be comfortable for us to live on. Without this greenhouse effect, life, at –18°C, would not be possible.

If the atmosphere absorbed and re–emitted all the IR passing through it this would represent the maximum greenhouse heating the planet could experience. The appendix to this chapter calculates this and shows that it would result in a temperature of the Earth's surface of 303K (30Deg C). This would be a rise of 15 °C over the present situation. Obviously an intolerable temperature, but at least it is interesting to know that there is a limit to this effect.

Mankind has been putting compounds into the atmosphere which are greenhouse gases, and this will therefore have an effect on the earth's temperature. The question is how much will this anthropogenic interference with the composition of the atmosphere affect planet temperatures?

To answer this question we need to look more closely at the mechanism of atmospheric IR absorption. Which components are involved, and what is the effect of changing their concentrations?

2.4) Radiation Absorption and Emission by Gases

All matter consists of molecules and atoms, which are collections of electrons orbiting around nuclei, each with a specific energy level. It is possible for these electrons to change their orbitals to different energy levels if there is sufficient energy to do so. This depends on the general amount of energy around and that depends on its temperature. Quantum theory shows that energy levels are in discrete packets, or quanta, and so the energy changes can only be in specific steps. The electrons can only jump between distinct energy levels, and so all energy of absorptions and emissions have discrete levels. In terms of radiation, energy levels are represented by wave lengths, and so absorption and emissions are at specific wavelengths

If the systems are simple enough, then the radiation passing through these materials is absorbed in distinct lines in the spectra, and any energy emitted will also be in the form of specific spectral lines. If the systems are complex, such as solids, then there are so many lines that the spectrum looks continuous − hence we can talk about black body radiation being a continuous spectrum. Gases are relatively simple systems, and so here we do see all spectra in the form of lines.

Gases are simple molecules freely moving around with a limited number of ways of holding energy. They have simple molecular motion and internal energies which can change levels. The electrons can 'jump' from one orbital to another. Electrons can 'spin' in different ways, and chemical bonds can 'vibrate' to different degrees. All altered energy levels within the molecule have distinct values − quantum theory allows for only very specific energy changes. Each energy change produces a corresponding radiation frequency or spectral line, and so the spectrum of radiation emitted from a gas consists of a series of lines (spectral lines) see Figure 2.3

Since absorption spectra and emission spectra are complementary, Figure 2.3 also represents absorption spectra of radiation passing through a gas.

Not all gases absorb IR. Only those gases which have energy level jumps which are in the quantum range of IR will absorb IR. Many simple molecules, such as oxygen and nitrogen have no absorption spectra in the IR range. The more complicated the molecule, the more likely it is to have absorption spectra with the IR − CO_2, H_2O, CH_4, all have IR absorption spectra. Also many man−made chemicals which have now escaped into the atmosphere have IR absorption properties – the chlorofluorocarbons (CFCs), in particular. Gases in the atmosphere which have absorption in the IR range are called greenhouse gases – GHGs.

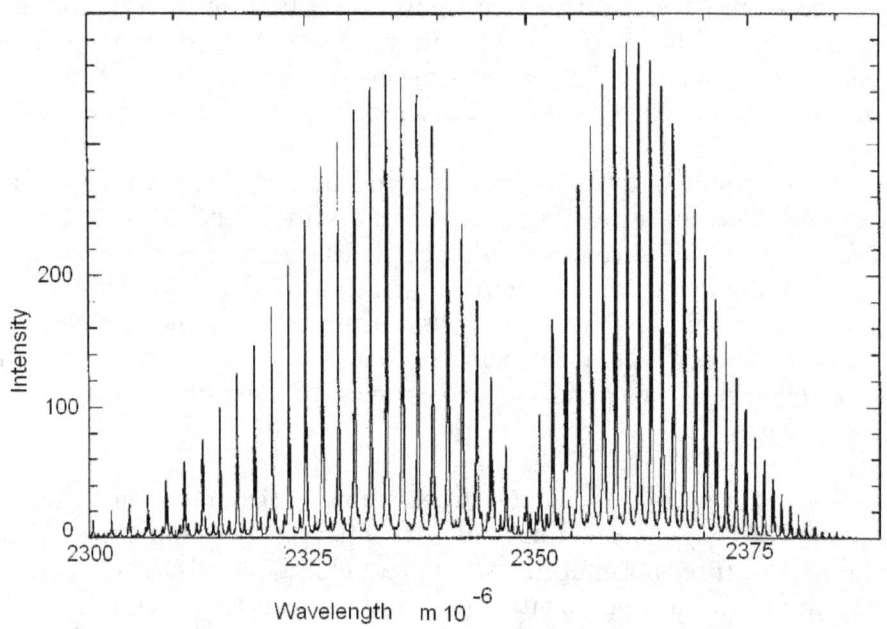

Figure 2.3 Part of the CO2 absorption spectrum,
Showing the spectral lines, and the band broadening, forming the 'wings' at the base of each line. The spectral lines themselves are so strong that higher concentrations cannot increase absorption, because these wavelengths have already been completely absorbed by very low CO2 concentrations. The wavelengths of the 'wings' at the base of the spectral lines however act as weak absorbers, and do absorb more as concentrations increase.

If a new GHG is introduced into the atmosphere it may absorb radiation at some new frequency which was previously not absorbed– often referred to as a 'window' in the IR spectrum. It will therefore result in a surface temperature rise in response to the downwards emitted radiation at the new frequency. This is the simplest absorption situation to explain – as previously that radiation got through unhindered, but with the new gas, these frequencies are absorbed and re–emitted in all directions.

If the concentration of an existing GHG – CO_2, H_2O or CH_4, for instance, is increased, then we need to be able to calculate what effect this has on the absorption coefficient. This is no longer straightforward as we have to look more closely at the effect of concentration on absorption.

Absorption of radiation is a very useful method of measuring concentration – from analyzing the composition of the atmospheres of other planets to the analysis of flue gas from boilers. Under the right conditions the absorption can be directly proportional to the gas concentration, which is good for the design of concentration analyzers.

Absorption occurs by photons of radiation energy with the right frequency, colliding with a gas molecule requiring that frequency to be able to absorb in a definite energy change. Double the number of molecules in a given space (doubling the concentration), and the number colliding is doubled, doubling the height of the absorption peak.

But now look at a significant depth of gas – not just a thin layer to which the above description applies. We can consider a thick layer of gas to be a series of slices, so now the absorption occurs over this series of 'slices'. If the absorption per slice is small and there are not many slices then there is still a good linear correlation between absorption and concentration. But as the amount absorbed increases, there are progressively fewer photons to be absorbed, and so the absorption is no longer linear. In the extreme, when the absorption is very strong, increasing the concentration has no effect on the absorption, because there are no more photons to be absorbed – the situation is often called 'saturated'. So when we increase concentrations, it is necessary to know what part of the absorption curve we are working with – is the absorption strong or weak?

The relationship between absorption and concentration is defined by Beer's law. This is developed by integrating the absorption over a distinct thickness, and the result is an exponential relationship. This is a very common form of relationship in the natural world, called first order, and is seen as defining   many events – from chemical reactions to radioactive decay.

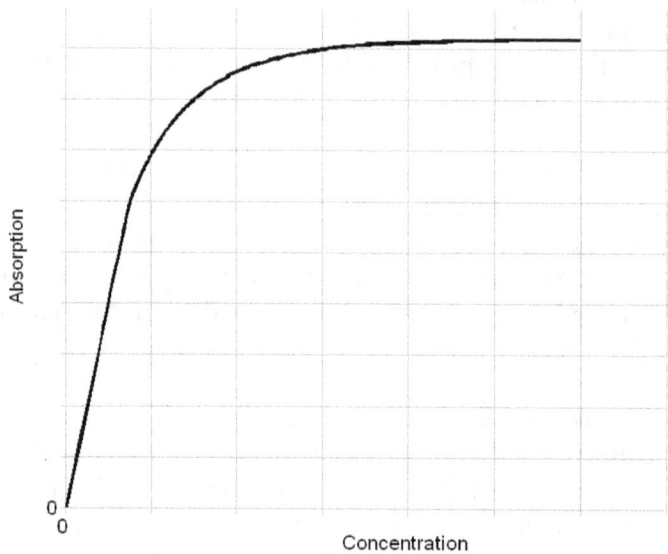

Figure 2.4   Beers law absorption curve
The exponential first order   curve shows the absorption to be linear at low absorptions, but curving to a plateau at ' saturation'

The mathematics of Beer's law is given in the Chapter 2 Appendix, where there are also some equations developed specifically for CO_2 . This is necessary to enable the modeling of global warming which will be considered in later chapters.

Figure 2.4 shows this relationship diagrammatically.

So whenever we need to look at the effect of concentration change on absorption, we must establish on what part of the Beer curve we are operating. This is particularly relevant when investigating the effect of increasing CO_2 atmospheric concentration on its greenhouse effect.

2.5) IR absorption by the Atmosphere, Modeling and Forcing

We have shown, with a very simple one layer atmosphere that there is greenhouse effect. But we know the atmosphere is not all at an even temperature, and so our greenhouse gas model needs refining.

The concepts behind the IR absorption of gases is very complex, and to fully understand the subject it must be treated in considerable detail. This detail we have separated into the Chapter 2 Appendix, as it is fundamentally important to have an understanding of it to be able to discuss global warming and the greenhouse effect. In this main body of the chapter we will only summarize the conclusion developed in detail in the Appendix.

The first stage is to accept that the absorption is over a depth of gas, requiring considering the atmosphere as a number of 'slices'. The absorption occurs in spectral bands and so the calculation must be made for each band, requiring thousands of calculations to cover the whole spectrum. Most of the spectral lines are strong absorbers and so will 'saturate', but because of a phenomena known as band broadening, which can be seen on Figure 2.3, there are low absorption segments at the foot of each spectral band, known as 'wings' .It is absorption by these wings, which do not saturate, which are affected by concentration. The whole problem becomes one of intense calculation, but spectroscopy is a very mature and well developed subject, so the calculation can be trusted.

Forcing

As this chapter Appendix detailed, the analysis of the passage of IR through the atmosphere is so complex it has to be handled by mathematical models. These models calculate the amount of radiation originating from the planet surface which leaves the upper atmosphere (Top of Atmosphere – TOA). Different compositions of the atmosphere, caused by anthropogenic CO_2, cause the radiation leaving to be changed, because of more absorption. This difference is expressed in W/m^2, and is called the **'forcing'** caused by the change in atmospheric composition – because it forces the planet to increase its temperature to compensate by this amount. All global warming is because of this 'forcing' and it will be a term regularly used through this text.

2.6) Results of IR Absorption modeling using MODTRAN4

Spectral models are available and one, MODTRAN4 (Berk, 1999) is used in the chapter appendix to investigate the effective absorption of the GHGs in our atmosphere. The results of this model are quoted in terms of 'forcing', the reduced emitted IR flow at the top of atmosphere (TOA) in energy units W/m^2.

The use of this model is demonstrated, and a table of the consequences of the forcings produced by each of the GHGs is given. The concentration effects for CO_2 and water are investigated by the model, and the effect of concentration changes in these components is estimated.

Finally, the Appendix calculates the temperature rise required to compensate for these forcings using Stefan's Law. To compensate for the forcing, global temperatures must rise 1 °C for every 3.4 W/m^2 of forcing.

The overriding impression from this quite detailed study is that the 'global warming' temperature rises caused by the various components is surprisingly small.

2.7) H2O Absorption and Feedback

A commonly used concept in climatology is 'Feedback', where a secondary event occurs which magnifies (or reduces) an initial event. In particular, when a temperature rise initiates an event which causes a further temperature rise. This is exactly the case of the forcing due to water vapor. Chapter 2 Appendix shows a 1 °C rise in temperature causes a 6% increase in water vapor pressure, which causes a 1.0 W/m^2 change in forcing, which causes a further 1.0/3.4 = 0.29 °C rise in planet surface temperature.

Feedback (f) is defined as the secondary rise in temperature caused by a 1 °C primary temperature change, so, for water vapor:–

$$f_{H2O} = 0.29.$$

This will be needed in later chapters.

2.8) Conclusions

In this chapter we have defined the greenhouse effect and shown that it is essential for tolerable temperature levels on the planet. We have also shown that as gas concentrations increase there is an increased greenhouse effect. The man–made components are particularly dangerous because they absorb IR from spectral regions 'windows in the spectrum' where there was previously no absorption. Increases in concentration of existing GHGs are relatively less important because they increase absorption in areas where there is already strong absorption, so their effect is relatively small.

It appears that, based just on radiation considerations, the planet surface temperature should increase by about 1.1°C from a doubling the CO_2 concentration and that the anthropogenic GHGs other than CO_2 have increased the temperature by 0.26°C based on the 2010 concentrations. It also shows that the water has a greenhouse effect, increasing by about 0.3°C for every 1°C increase from other causes.

These examples show that the temperature of the surface of the planet will be affected by changes in atmospheric composition. The next step is to estimate what will be likely future concentrations of the GHG gases, particularly CO_2, that the atmosphere of the planet is likely to experience. Later on we must also investigate additional effects that occur which may augment this temperature rise from radiation absorption and re–emission alone – the possible 'feedback' effects.

Appendix to Chapter 2 – Going into Detail

To determine the equilibrium temperature to be expected on our planet we can look to Stefan's Law, which is summarized by equation 2.1 .

The planet is heated by the radiation falling on it from the Sun. The intensity of the radiation falling perpendicular to the Earth has been measured by satellite to be 1367 W/m^2 (S). The amount of radiation falling perpendicular on the Earth is therefore equal to an area of a circle the diameter of the Earth (D m), (the Earths shadow), and so the total radiation received by the planet is

$$1367 \text{ x } \pi \text{ x } D^2 /4 \text{ W}$$

The total surface area of the planet is $4 \text{ x } \pi \text{ } D^2/4$, so the mean radiation falling on one m^2 of the planet surface is

$$\frac{1366 \text{ x } \pi \text{ x } D^2/4}{4 \text{ x } \pi \text{ } D^2/4} = 1367/4 = S/4 = 341 \text{ W/m}^2$$

The average 'albedo' , the fraction of radiation reflected from the surface of the planet, is 0.3, so the energy reaching the surface of the earth averages out at :

$$341 \text{ x } (1.0 - 0.3) = 239 \text{ W/m}^2$$

This is the amount of energy that the planet must lose through emitting IR radiation, on average, per square meter of surface for the earth to remain in energy balance.

Using the Stefan–Boltzmann law, given by equation 2.1 we can calculate that the mean planet temperature to lose this energy is:

$$T^4 = I / \sigma \quad \text{ or } \quad T = (I / \sigma)^{1/4}$$
$$= (239 / 5.67 \times 10^{-8})^{1/4} = 255\text{K} (-15°\text{C})$$

Clearly this mean planet temperature is much lower than the real temperature which is generally thought to be about 288 K, (15°C). There is a discrepancy of 33 K.

The Greenhouse Effect

The reason for this discrepancy of 33K is that the planet is covered by a layer of gas, – our atmosphere. If our atmosphere let all the IR radiation from black body earth through into space, then the temperature would be 255K, as calculated. However, the atmosphere contains a number of gases, which absorb some IR. We need to look closely at the effect of this partial absorption of IR.

Let us now consider a planet surrounded by an atmosphere. To make it simple, assume that this atmosphere is a well mixed layer of gas. IR from the earth's surface passes through this layer, and a certain fraction (ε) is absorbed.

Everything, including gases, that absorbs radiation, emits radiation and with emission coefficients (ε) that are the same as the absorption coefficients. A basis of all radiation theory is that the emission pattern corresponds exactly to the absorption patterns. A body at a constant temperature is receiving radiation from its surroundings, but is emitting exactly the same amount of radiation – otherwise it would not remain at a constant temperature.

But the difference between absorption and emission is that the absorbed radiation is in only one direction – the radiation leaving the earth – but the emission is in two directions – from both the top and the bottom of the gas layer. So the earth's surface emits radiation corresponding to the surface temperature (T_s) in an upward direction, which is absorbed by the layer of atmosphere which re–emits radiation both upwards and downwards corresponding to the temperature of the atmosphere (T_a). The final, important atmospheric temperature is that at the upper extreme of the atmosphere, denoted 'Top of Atmosphere' (TOA).

Figure 2.2, in the main body of the Chapter, displays the radiation flows involved.

Chapter 2   Planet Temperature and Greenhouse Gases

The whole system must be in heat balance, both across the planet, and also across the atmosphere. So we can say the following:

Incoming radiation = $S/4$

Reflected from surface, albedo α,   = $\alpha \, S/4$

IR leaving earth's surface, temperature Ts = $\sigma \, Ts^4$

IR absorbed by Greenhouse gas   = $\varepsilon \, \sigma \, Ts^4$

IR from surface leaving planet unabsorbed = $(1 - \varepsilon) \, \sigma \, Ts^4$

IR emitted by greenhouse gas, temp Ta = $\varepsilon \, \sigma \, Ta^4$,
$\qquad\qquad\qquad\qquad$ both upward and downward.

We can now write 3 equations for the planet atmosphere, the planet in steady state, for the planet as a whole, and the planet surface, all being in heat balance.

Firstly, we can write a heat balance for the atmosphere layer– the heat absorbed by the GHGs must equal the heat emitted in two directions by them:

$$2\varepsilon \, \sigma \, Ta^4 = \varepsilon \, \sigma \, Ts^4 \qquad\qquad (2.2)$$

or

$$Ta = Ts/2^{1/4}$$

which shows that the temperature of the atmosphere will be lower than the surface temperature due to the greenhouse effect.

We can write a heat balance for the planet as a whole:– the energy entering from the sun must equal energy leaving at the top of the atmosphere. This energy comes from the unabsorbed radiation from the Earth's surface and the re–emitted absorbed GHG radiation emitted in the upward direction:

$$(1 - \alpha) \, S/4 = (1 - \varepsilon) \, \sigma \, Ts^4 + \varepsilon \, \sigma \, Ta^4 \qquad (2.3)$$

34

We can also write a heat balance for the planet surface. The energy received from the Sun, plus the reflected GHG downward emitted radiation must equal that energy leaving the surface:

$$(1 - \alpha)\, S/4 + \varepsilon\, \sigma\, Ta^4 = \sigma\, Ts^4 \qquad (2.4)$$

Now let us look at the equilibrium surface temperature, receiving this extra IR emitted down from the atmosphere.

From equations 2.3 and 2.4 we can say:

$$(1 - \alpha)\, S/4 + \varepsilon\, \sigma\, Ts^4 /2 = \sigma\, Ts^4$$

$$(1 - \varepsilon /2)\, \sigma\, Ts^4 = (1 - \alpha)\, S/4$$

so;

$$Ts^4 = \frac{(1 - \alpha)\, S/4}{(1 - \varepsilon /2)\, \sigma} \qquad (2.5)$$

Using the numerical values already mentioned in this chapter we could calculate the planet mean surface temperature when covered by a layer of green house gas, except that we do not know the absorption coefficient (ε) for the atmosphere. If we take the accepted temperature of the Earth's surface as 288 K, and work backwards with equation 2.5, we can find by trial and error a value of 0.78 as the absorption coefficient for the atmosphere.

This means that our atmosphere absorbs 78% of the IR from the earth, and re-emits this in both directions. It is this re-emission in the downward direction which has given rise to the name of the phenomena as the 'greenhouse effect'. The planet has to thank the greenhouse effect for making the temperature high enough for it to be comfortable for us to live on. Without this greenhouse effect life would not be able to exist.

If the atmosphere absorbed and re-emitted all the IR passing through it this would represent the maximum greenhouse heating the planet could experience. Setting $\varepsilon = 1.0$ into our equations give a temperature of the Earth surface of 303K (30°C). This would be a rise of 15 K over the present situation. Obviously intolerable, but at least it is interesting to know the limit.

To simulate the case of no absorption in the atmosphere, we can set ε to 0 in equation 2.5 and we get a surface temperature of 255 K, agreeing with the analysis we made at the beginning of this section.

This analysis has been very simple. We now need to look more closely at the mechanism of this atmospheric absorption. Which components are involved, and what is the effect of changing their composition? Even later, we will discuss the problem that the atmosphere is not one isothermal layer, but 100 km thick with its own temperature profiles, which alter the mathematics, requiring a much more complicated analysis to be realistic.

Beers Law

The main text explains that it is necessary to know the relationship between absorption and concentration, and this is defined by Beer's Law. In this Appendix we develop the equations for the law, and fit a specific set of parameters to represent CO_2 in our atmosphere.

Consider a slab of gas being irradiated with IR, with the gas being able to absorb this radiation.
Given the intensity of the IR is I , the slab thickness being z, with the absorption per unit thickness being β c, where c is the concentration of absorbing gas and β an absorption coefficient.

The absorption over a thin slice of the slab dz , is dI and, being a first order process, the relationship is:

$$-dI = \beta \, c \, I \, dz$$

integrating over the thickness of the slab Z , with a starting intensity Io gives:

$$-\int dI/I \; = \; \beta \, c \int dz$$

$$\ln I/Io \; = -\beta \, c \, z$$

or
$$I \; = \; Io \; e^{-\beta \, cz}$$

in terms of absorption not intensity, $A = Io - I$ and the equation becomes :

$$\mathbf{A \; = \; Io \, (1 - e^{-\beta \, cz})}$$

This is a formulation of Beer's law, whcih shows an exponential (sometmes referred to as logarithmetic) relationship between concentration and absorption

Figure 2.4, in the main body of the chapter, shows this relationship diagrammatically.

So whenever we need to look at the effect of concentration change on absorption, we must establish on what part of the beer curve we are operating. This is particularly relevant when investigating the effect of increasing CO_2 atmospheric concentration on its greenhouse effect.

A detailed Analysis of IR absorption by the Atmosphere

We have shown, with a very simple one–layer atmosphere that there is greenhouse effect. But we know the atmosphere is not all at an even temperature, and so our greenhouse gas model needs refining.

Our single layer isothermal atmosphere must be changed to incorporate the fact that the atmosphere is not isothermal. This is done by constructing a model with more than one layer. A model with

10 layers is common. The temperature in the atmosphere reduces with height. One reason for this is because the rising air reduces pressure, the volume increases, but the energy content remains essentially the same. Under these conditions the gas laws show that the gas temperature must reduce.   This reduction is called the adiabatic lapse rate, and it is 6.5 °C/km. This is the basic cooling rate as air rises until there are other factors involved which will change the heat content – eg condensation of water, convection, or IR absorption.

Now we have an atmosphere with a number of layers, we can make a heat balance for each layer to determine its equilibrium temperature. Whereas for our simple model the heat balance contained only 3 terms, as shown in equation 2.4, for our new atmospheric multilayer model each layer has 4 IR streams (the fourth being a reflected IR stream from the layer above), and a temperature change due to the lapse rate and changes from condensation. The solution of this multilayer model is more complex because there are more layer temperatures to solve for, and the reflected radiation from the layers above require an iterative equation solver to be used. This model is however a much closer description of the physical system.

We still have the second problem of defining the absorption coefficient built up from the composition of the atmosphere. This means taking all the spectra, from all the components, and calculating their absorption coefficients for all their spectral lines. This can be done because the spectral lines of all compounds have been measured and many have also been calculated from quantum theory from their molecular structure. By using data bases for the spectral lines, it is possible to calculate the absorption coefficients for each line of each spectrum.. The calculation of these spectral lines is very calculation intensive. because there are many thousands of lines to consider. There are various mathematical techniques for reducing the computational load such as LBL, (Line–by–line), NBM(Narrow band), BBM (Broad band), models. These are computational possibilities which do not alter our discussion but are often referred to in modeling text.

Because most spectral lines of GHGs are strong, this means that increases in concentration should not be expected to cause significant increases in absorption because they represent working on the plateau of the Beer curve where absorption is complete. Only weak lines, being on the early part of the Beer curve, increase their absorption as concentration increases over the 100 km depth of the atmosphere.

Many global warming deniers point out that the strong GHGs are not going to be any more detrimental at higher concentrations because they have already absorbed all the IR they can – they are saturated in the Beer's law sense.

But strong lines, when looked at closely are not straight lines, but are spread out at their base – see Figure 2.3. This spreading out at the base is called 'band broadening' and it is due to pressure or Doppler effects, which are second order effects discovered by spectroscopists. These spreaded bands, or 'wings' are equivalent to bands of low absorption and their absorption is affected by concentration. So it is the absorption in these wings where changes in concentration of existing GHGs may affect thermal balances.

This means that even strongly absorbing species such as CO_2 and H_2O do absorb more IR at higher concentrations from the 'wings' of the spectra. But all can be calculated, spectroscopy is a very mature subject, being used by astronomers, space scientists, physicists chemists and analysts, so plenty of data is available and proven calculation methods abound.

The matter is so complex that we must leave it to the computational ability of the spectroscopists to define the final effect to be obtained from various atmospheric compositions. Their methods are very accurate; figure 2.5 shows how well they can predict a measured spectrum. This figure shows a spectrum observed by satellite, compared with the calculated spectrum using theoretical and stored data bases for the spectra of the components, the agreement is remarkable (Chen, 2007).

Results on Radiative Forcing from MODTRAN4

There have been a number of radiative models of the atmosphere, following the concepts of the previous section. – ie multi–layer, with detailed spectral line analysis. One particular model MODTRAN 4

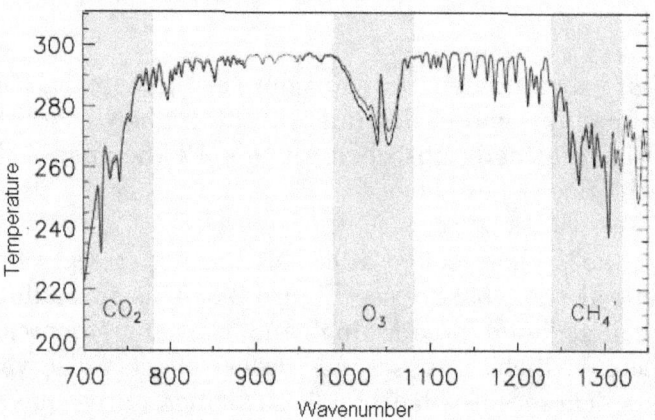

Figure 2.5 Accuracy of Calculation of Spectra

Comparison of the spectrum leaving the top of the atmosphere (TOA), measured and calculated. The agreement between measured IR spectrum (thin) and the calculated spectrum (thick) is remarkable, being virtually superimposed,except for the region around 1000 wave number. Spectroscopy is a very well understood subject, and predictions of IR absorption by gases is very accurate.

(Berk,1999) is a version of the U.S. Air Force atmospheric transmission, radiance and flux model which has been developed for aerial IR monitoring of activities on the ground,

The coding is publicly available and has been prepared for use via the internet;

http://geoflop.uchicago.edu/forecast/docs/Projects/modtran.orig.html

to demonstrate the effect of various atmospheres on the IR leaving the planet. The coding requires a defined surface temperature, and atmospheric composition, and a surface water partial pressure. Calculated are the temperature and pressure profiles in the atmosphere, the composition profile of the H_2O and the total IR leaving the planet, going into space. This can be run to determine the effect of CO_2 or CH_4 concentrations on the emitted radiation, and the effect of water.

To start the investigation with MODTRAN4 it is necessary to choose a set of input data as a standard case. To compare with the simple calculation done in the earlier sections of this chapter, conditions were chosen which gave a planet mean temperature and outgoing radiation close to the example made previously. The best matching set was taken as 280ppm CO_2, H_2O vapour pressure taken as saturated at the surface temperature and a surface temperature selected as 287K. The MODTRAN code gives an outgoing radiation of 252.4 W/m^2 for this situation, which represents an albedo of 0,28 to balance with the incoming average solar radiation.

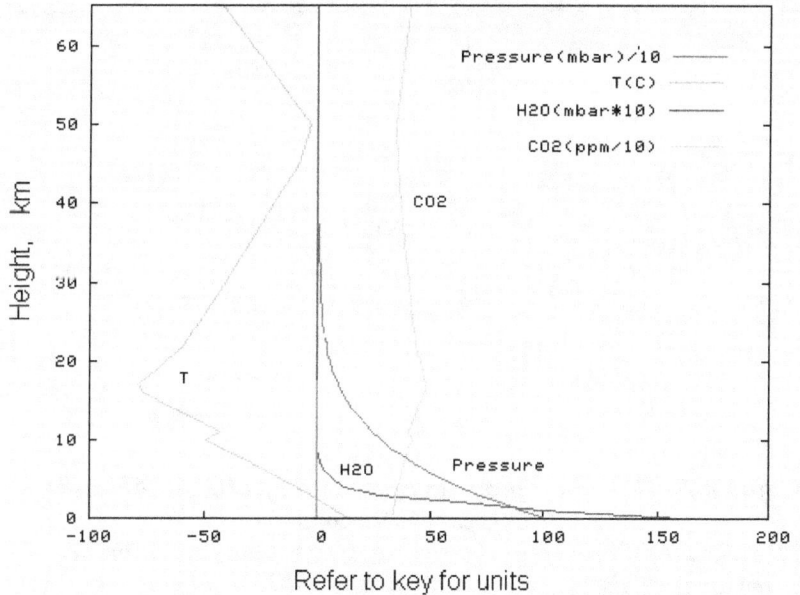

Refer to key for units

Figure 2.6   Model MODTRAN calculated Atmospheric Profiles
Temperature *(T) falls with height according to the lapse rate until reaching 17 km where it increases – albeit at very low pressures.*
   Pressure *falls asymptotically over 30 km*
   Water *vapor pressure falls very rapidly to be almost zero at 8 km*
   CO₂ *concentrations are fairly constant over the whole height*

Taking this set of input data the computer results gave the profile shown in figure 2.6. Of interest in this profile is the resulting complex temperature profile which is a mixture of adiabatic lapse rate plus the radiation balancing equations equivalent to equation 2.4, but with more terms to include radiation from the layer above. The result is the complex inversion profile. The temperature profile reached a low temperature of −80 C and this is evident in the spectrum produced The water profile quickly becomes very low in concentration, as the upper atmosphere holds very little water. Again the spectrum will show that the water spectrum has a higher temperature than CO_2 because the IR emitted from the water is in a layer of atmosphere at a higher temperature (closer to the surface) than the emitting layer for CO_2

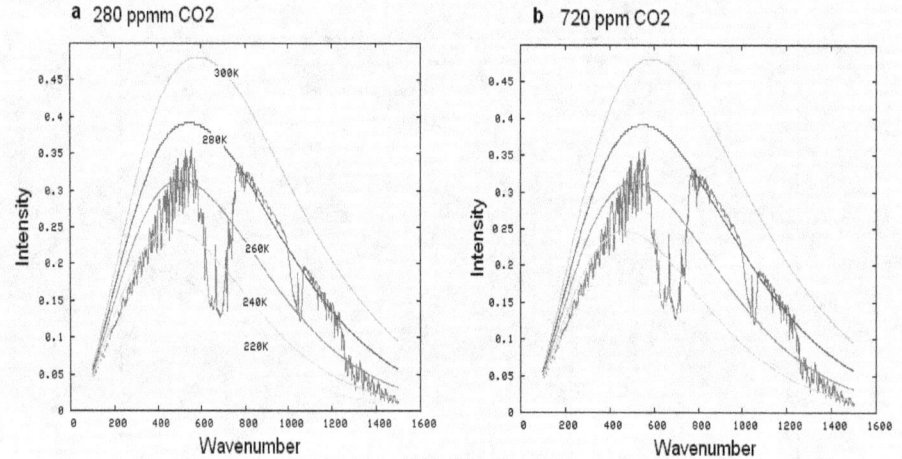

Figure 2.7: TOA IR Spectrum calculated by MODTRAN at two
CO_2 concentrations

The small difference in the two spectra are due to differences in the 'wings'. This shows itself in the full spectrum as small differences at the edges of the CO_2 band. The 720ppm CO_2 band is distinctly wider.

Figure 2.7a shows the spectrum for this standard case. The continuous lines are the black body equivalent emission spectra at different surface temperatures. From this it can be seen that where no absorption takes place the spectra is up to that surface

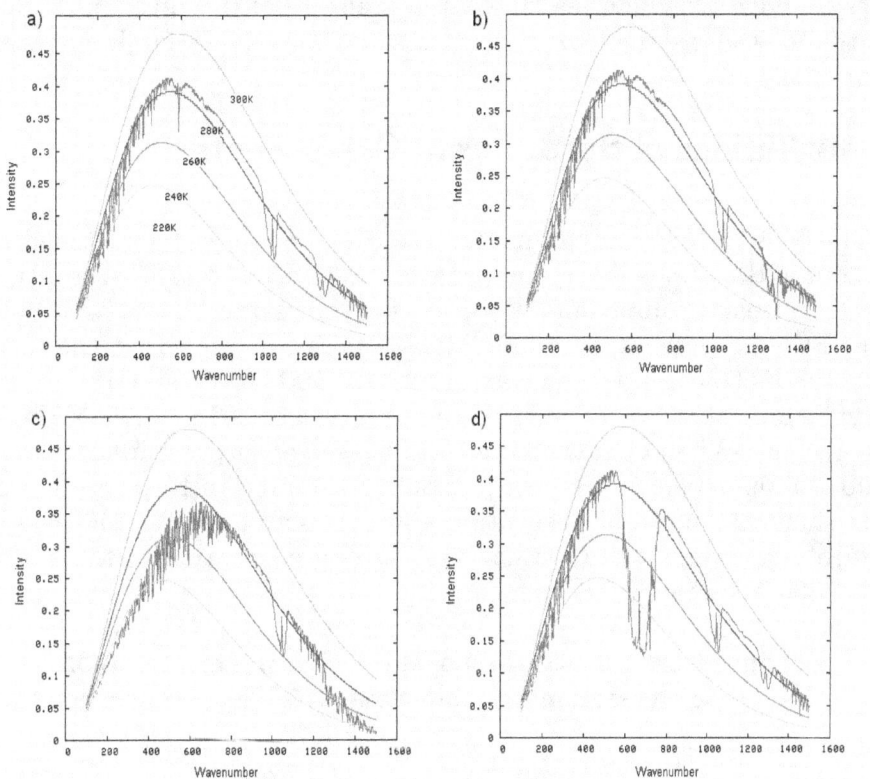

Figure 2.8: TOA IR Spectrum calculated by MODTRAN for different GHG gases

a) an atmosphere containing the CFCs and ozone as the only GHGs. The difference between the smooth curves and the spectra, represent the absorption by the GHGs. The ground temperature can be seen to be about 283 K from the position of the spectrum against these black body curves.

b) adding just CH_4 to spectrum a shows the absorption of $CH4$, which is in the 1300 – 1400 wave number range

c) adding just $H2O$ to spectrum a shows the strong absorption of $H2O$, which is in the 200 – 700 and 1300 – 1500 wave number range. The absorbed spectra is re–emitted at temperatures of 240 – 260K – higher in the atmosphere

d) adding just $CO2$ to spectrum a shows the strong central $CO2$ band in the 600 – 800 wave number range. The absorbed spectra is re–emitted at a temperature of 220K, – much higher in the atmosphere.

temperature of 287K. Where absorption does take place, the spectra lines up with those curves where the majority of the final emissions occur – of the order of 220K for CO_2 and 240–260K for H_2O – see figure 2.8.

Contributions of the different Components

The MODTRAN coding enables the components to be switched in and out of the atmospheric composition, so illustrating the absorption spectra of each component. Figure 2.8a shows the case where only the CFCs are present, at 1990 values – it can be seen that the CFC have distinct absorption at wave numbers 500, 1000, and 1300. The curve is almost up to the surface temperature of 287K between these absorptions. When CH_4 is switched in we see absorption in the range 1200 – 1400 as shown by Figure 2.8b: Switching in H_2O in place of CH_4 shows 2 ranges of absorption 100 – 700 and 1200 – 1400 – see Figure 2.8c;.finally the CO_2 has very strong absorption in the range 500 – 800 see Figure 2.8d

Table 2.1
Forcings from trace anthropogenic GHGs other than CO_2
due to increases from pre–industrial to 1995 concentrations

Compound	Forcing W/m^2
CFC–11	0.09
CFC–12	0.20
Total Other CFCs	0.10
CH4 *	0.62
N2O,	0.15
Total	**1.17**

*Note that the CH_4 forcing is caused by 1700ppm, the pre–industrial value was stable at 700ppm, hence 40% of this CH_4 forcing (0.26W.m^2) could be considered background, leaving the total anthropogenic GHG forcing excluding CO_2 in 1995, referred to in later chapters as F$_{other}$, to be 1.17–0.26 = **0.91 W/m^2***

Between these absorptions there is a region where very little absorption takes place, namely 800 – 1250. This is called a 'window' through which much of the IR escapes the planet. Components absorbing in this region are of particular concern because they are in this window, The CFC are absorbed in this region with a strong peak at 1000 – 1050, so are very effective GHGs even at parts per trillion (ppt). The forcings from the anthropogenic Chlorofluorocarbons (CFCs) is considerable. Table 2.1 shows estimates of these forcings (Myhre 1998) for concentrations measured in 1990s

Effect of Changes of Concentration

The effect of changing CO_2 concentrations can be simulated by MODTRAN by running at 280 and 560ppm and noting the difference in the out–going radiation. This difference is the radiative forcing due to the increase in concentration. The result of this investigation is a value of 2.5 W/m^2 for a doubling of the CO_2. The effect of water can be determined by changing the partial pressure of the water at the surface equivalent to 1K and noting the change in outgoing radiation. This simulation shows that a 1K change in surface temperature, which changes the water vapor pressure by 6% will produce a forcing of 1.0W/m^2. This shows that water is by far the most important greenhouse gas, and is far from saturated, because the 6% change of water concentration has the same forcing as a 40% change in CO_2. A similar exercise can be made with CH_4; by changing from the standard 1.7 ppm to 3.4ppm produces a forcing of 0.6 W/m^2 – somewhat less significant than CO_2.

The results from this simple MODTRAN study suggest doubling the CO_2 will produce a radiative forcing on 2.5W/m^2, whereas the most accepted value in the literature, and endorsed by the IPCC is 3.7W/m^2. If MODTRAN is run with its default data, which represent 1990 conditions, then doubling from the 375 to 750 PPM give a forcing of 3.1 W/m^2 – which is closer to the 3.7 generally accepted. These calculations are only for illustrative purposes, and so the difference between 3.1 and 3.7 is not significant in our discussion. But it does show how complex the subject can become – the model stipulates tropical conditions, over the ocean, no clouds, the ground temperature is not the planet mean temperature but higher (the fourth power makes arithmetic means inappropriate). A particular water model is assumed, and the coding is being used by a third party not aware of its limitations and constraints.

CO_2 Absorption and Beer's Law

Beer's law shows that the absorption/ concentration relationship has a first order exponential form, and this can be confirmed using MODTRAN as follows: By taking the standard case, and simply altering the CO_2 concentration over a concentration range from 280ppm to 1500ppm we can plot the relationship between absorption and concentration. Figure 2.9 shows the relationship between outgoing radiation, (expressed as the radiative forcing by CO_2 concentrations above 280ppm) and concentration. As expected from Beer's Law, this shows an exponential decay to saturation.

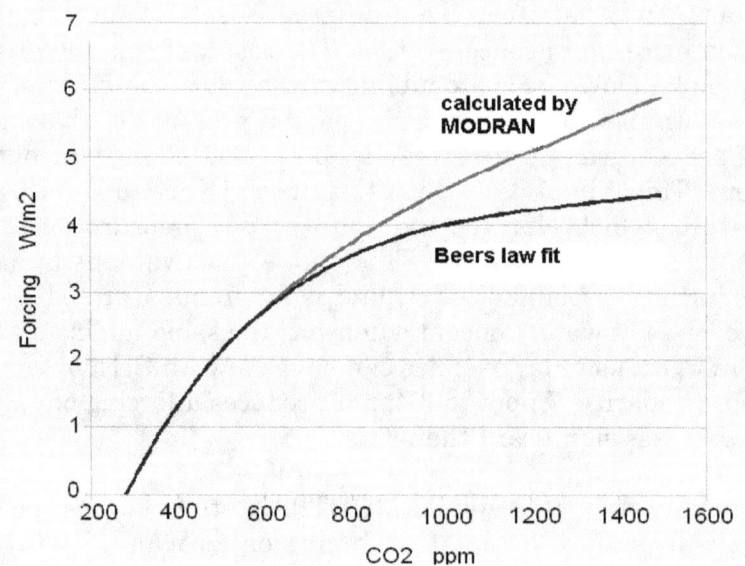

Figure 2.9 Radiative forcing increase produced by CO_2 concentration increases above 280ppm

The Beer equation represents just one spectral line, whereas the CO_2 absorption contains thousands of spectral lines with 'wings'. This means the equation can satisfactorily represent the results over part of the concentration range, but not over the whole range. The equation becomes unsuitable for concentrations over 1000 ppm.

This curve can be fitted by an equation with the same form as Beer's law, so that the relationship can be used in prediction and modeling. The forcing above 280ppm CO_2 created by the actual CO_2 concentration can be represented by the equation:

$$F_{CO2} = B(1 - e^{-\beta z\,(p-280)})$$

and this can be fitted to the MODTRAN simulation as shown in curve figure 2.9 This fit has been obtained with the parameter values:

$$B = 4.60, \quad and \quad \beta z = 0.0028$$

This equation with its parameters can be used in predicting the effect of future CO_2 concentrations and will be used in later chapters. This relationship means that CO_2 becomes progressively less of a greenhouse gas at higher concentrations.

It is interesting to note that this equation fits concentration below 550pm very well, but then there is divergence between the MODTRAN results and the single Beer equation model. This can be explained by realizing that the Beer equation is for single spectral line, which will asymptotes to saturation, whereas the MODTRAN4 is simulating the real world, which is a whole range of spectral lines of different intensity, which will asymptote at different concentrations. Hence the clear asymptote of the theoretical models is diffused by the number of different lines. The simple model is fine for concentrations up to 550ppm, beyond which there will be progressively more error. To overcome this we would need a more complex equation with a number of terms and many more parameters to represent the different strengths of spectral line. The very simple model is very useful, but it has its limitations, but still can represent our system accurately enough over a useful range. Where the simple model diverges from the real data, gives us a clue as to the complexities of the real system. It is worth laboring this point as it well illustrates that the use of fitted semi–theoretical models, based on theoretically justified equations, can be very useful for prediction. This will be important in later chapters.

Relating Surface Temperature to Forcings

Converting changes in surface temperature to maintain the planet radiation heat balance is possible using MODTRAN by changing the surface temperature until the standard 252.4 W/m^2 is achieved. The change in temperature indicates the change in mean planet temperature to counter–act this greenhouse forcing. This analysis with MODTRAN shows that a 1K surface temperature is equivalent to 3.4 W/m^2 – a change of 1.35%.

This information can be compared with the Stefan fourth power law which says that a radiation change by 1K surface temperature would be a

$$(288/287)^4 = 1.0140 \text{ or } 1.40\%$$

These MODTRAN4 results would suggest that the doubling of CO_2 causing a forcing of 3.1 will cause a temperature rise of 0.9 K (staying with the MODTRAN results). or using the normally accepted 3.7 W/m^2 will give a 1.1 K rise in temperature.

The increase of 0.9K will increase water vapour pressures by 5.4% . This gives water vapour forcing of 0.9W/m^2 , which together with the CO_2 forcing produces an increase of 4.0 W/m^2 (4.6W/m^2) for a doubling of CO_2 including the effect of increased radiative forcing from water. This requires a surface temperature increase of 1.2K (1.4K) to maintain planet heat balance.

This is looking only at the radiative effects of CO_2 and the 'feedback' from the radiative effect from the increased water vapor presence. It does not look in detail at any other effects – clouds, geographic differences, ice caps etc, which can be investigated using more complete multi–dimensional global models, and will be the subject of later chapters.

References to Chapter 2

Berk A., G. P. Anderson, L. S. Bernstein, P. K. Acharya, H. Dothe, M. W. Matthew, S. M. Adler–Golden, J. H. Chetwyn, Jr, S. C. Richtsmeier, B. Pukall, C. L. Allred, L. S. Jeong, and M. L. Hoke, *MODTRAN4, Radiative Transfer Modeling for Atmospheric Correction,* SPIE Proceeding, Optical Spectroscopic Techniques and Instrumentation for Atmospheric and Space Research III, Volume 3756, July 1999, (www.spectral.com/pdf/sr116.pdf)

Chen, C., Harries, J., Brindley, H., Ringer, M., *Spectral signatures of Climate Change in the Earth's' infrared spectrum between 1970 and 2006,* Joint 2007 EUMETSAT Meteorological Satellite Conference and the 15th Satellite Meteorology Oceanography Conference of the American Meteorological Society (2007)

IPCC, Working Group I, to the Fourth Assessment Report of the Intergovernmental Panel on Climate Change, *The Physical Science Basis,* 2007 Solomon, S., D. Qin, M. Manning, Z. Chen, M. Marquis, K.B. Averyt, M. Tignor and H.L. Miller (eds.)Cambridge University Press,

Myhre G., IIighwood E. J., Shine K. P.,Stordal F., 1998, *New estimates of radiative forcings due to well mixed greenhouse gases,* Geophysical Research Letters 25, 14, 2715 – 2718, 1998

Further Reading

Modest, M. F., *Radiative Heat Transfer ,* Academic Press, 2003

Ramanathan V., *Trace–gas Greenhouse Effect and Global Warming,* Ambrio, 27, 3, 1998

Chapter 2 Planet Temperature and Greenhouse Gases

Chapter 3   The Role of the Oceans

3.1) The Global Carbon Balance

As the object of this book is to investigate the fate of CO_2 emitted from the burning of fossil fuels, we need to understand how much carbon is on our planet, and where it is located. Only when this is understood can we start considering what happens when we start adding more from fossil fuels.

So a good place to start is to look at the distribution of the present carbon inventory on the globe to see where it is all located.

One finds the majority of the carbon is present as calcium carbonate in sedimentary rocks, both beneath the sea and now above the sea, on land. This is by far the greatest quantity, being 99.998% of the total carbon on the globe. The next major repository of carbon are the oceans which hold carbon in the form of dissolved inorganic carbon (DIC), mainly carbonate ($CO_3"$) and bicarbonate (HCO_3') ions. Following this is terrestrial carbon in the form of biological carbonaceous material (C_{org}), from living matter. Finally there is the atmosphere with very little carbon – only about one fiftieth of the carbon in the oceans.   The amount in the atmosphere is so low because the mass of the atmosphere is relatively very small, and the CO_2 concentration is also low.

Table 3.1 summarizes these percentages, and figure 3.1 shows these quantities diagrammatically. The figure also shows the different transfers, or interactions, that occur between the various locations.

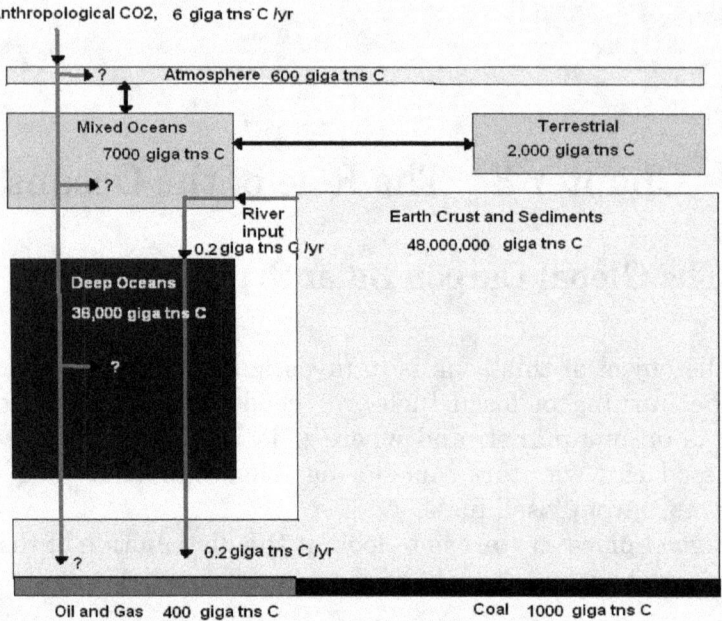

Figure 3.1   Global distribution of Carbon

There are 5 major sinks of carbon, with very different capacities. There is transfer between a number of these sinks, some where the transport must be equal in both directions, but the river transport to the sea is one–way. Before the onset of anthropological CO2, the system was in balance because the river flow was matched by the one–way burial of calcium carbonate on the sea bed.

Table 3.1 Global distribution of Carbon

Location	Carbon compound	Amount as Carbon
Atmosphere	CO_2	600 Gt C
Terrestrial	C_{org}	2000 Gt C
Surface oceans	DIC	7000 Gt C
Deep oceans	DIC	38000 Gt C
Sediments and crusts	$CaCO_3$	48×10^6 Gt C

Gt C = giga ton carbon $= 10^9$ ton = billion ton carbon

Terrestrial Sinks

There are interactions between the atmosphere and the terrestrial biological systems. These interactions are in both directions, and must be equal. Plants absorb CO_2 from the atmosphere, but eventually all this plant material returns as CO_2 to the atmosphere, either by respiration in animals or by burning, or by microbiological action of composting materials. It is unlikely that any long–term terrestrial sinks still exist, where material plant is fossilized by sinking into swamps, as was the case in the earlier history of the planet. If this was the case then the flows would not be equal in each direction, and the flow back into the atmosphere would be less.

A very important interaction to understand is between the oceans and the atmosphere. This is a normal physical interaction between a gas and liquid where the two phases strive to be in equilibrium. The CO_2 will also vaporize from the sea with water vapor and return to the oceans dissolved in the rain. Some CO_2–containing rain will have flowed over carbonaceous rocks returning with dissolved $CaCO_3$, thereby returning some CO_2 from these rocks back to the sea, as bicarbonates. This results in the flow of 0.2 Gt C/yr carbon from the sediments and crusts to the sea, as shown on figure 3.1.

Deposition on the Sea Bed

There is interaction between the sea and the sea bed, whereby the calcium carbonate from marine animals deposits on the sea bed on their death. This provides 0.2 Gt C/yr of new carbonaceous material deposit on the sea bed. So it is clear that the HCO_3' returning from the rivers is an essential part of the system. Without this there would be a slow depletion of calcium in the sea and this would be disastrous for marine life. The marine animals also contain carbonaceous cellular material (C_{org}) which make up the living organism. On death this material also sinks to the bottom of the sea. On land such biological material is decomposed by aerobic microbial action to produce CO_2. On the sea bottom, where there is very little oxygen, such biological material may not be oxidized, but attacked by anaerobic microbial action, with the eventual result being conversion to deposits of crude oil. The deposition of calcium carbonate on the sea–bed over the lifetime of the planet has been the source of all the carbon in the sediments and Earth's crust.

That the sea contains about 50 times as much CO_2 as the air suggests that the anthropological CO_2 should eventually divide up in the same ratio with 2% remaining in the atmosphere and 98% transferring into the sea. Once in the sea one might expect that it finally ends up as calcium carbonate on the sea bed, so providing a permanent sink for the anthropological CO_2.

This, we will see, is quite wrong.

The complex equilibrium between the various carbonates in the sea means that the more CO_2 dissolved in the sea, the less will CO_2 dissolve in the sea. This means a ratio of 10 rather than 50 is found to be a more realistic figure when comparing the distribution of additional CO_2 between air and seawater. This ratio is the *Revelle Factor* which will be explained in detail in the next section. As for calcium carbonate precipitating and so removing CO_2 out of harm's way, the opposite is in fact the case. In the sea, the bicarbonate holds twice as much CO_2 as the carbonate. To precipitate this bicarbonate as $CaCO_3$ will liberate half the CO_2 back into the sea. In fact the best way of ' neutralizing ' CO_2 is to dissolve calcium carbonate, which results in a higher level of calcium ions dissolved in the sea but which enables the sea to hold more CO_2.

3.2) The Chemistry

The oceans cover 70% of the earth's surface to an average depth of about 3.6 kilometers. They contain soluble material which has been collected from the land and atmosphere by the continuous cycle of water evaporation, rain, flow in rivers, and hence return to the sea. The seas contain soluble cations of sodium, calcium, magnesium, potassium and anions of chlorine, bromine and sulfate. On evaporation, the compound sodium chloride (common salt) is predominant, although other salts such as magnesium and calcium chlorides and sulphates will also be collected. The mean composition of sea water is shown in table 3.2.

Table 3.2
Major components and mean seawater composition
Standard mean chemical composition of seawater

Ion	g /kg	mol /kg
Na^+	10.7837	0.46906
Mg^{2+}	1.2837	0.05282
Ca^{2+}	0.4121	0.01028
K^+	0.3991	0.01021
Cl'	19.3524	0.54586
Br'	0.06730	0.00084
SO_4"	2.7123	0.02824
CO_2	0.00040	0.00001
HCO_3'	0.1080	0.00177
CO_3"	0.0156	0.00026
OH'	0.0002	0.00001
Sum	**35.2**	**1.12**

Sea water composition is defined by Salinity, DIC and Alkalinity.

Salinity *is the quantity of salts in the water, expressed as g/kg*
  (The water in table 3.2 has a salinity of 35.2 g/kg)

Dissolved inorganic Carbon (DIC) *is the sum of the carbonates, bicarbonates and dissolved* CO_2 , *expressed as mol/kg.*
So the water in table 3.2 has a DIC of

$$0.00177+0.00026+0.00001 = 0.000204 \ \ mol/kg$$

Alkalinity Sea water contains strong cations: Na^+, K^+, Mg^{2+}, and Ca^{2+}, and strong anions: Cl', Br' and SO_4'' and weak anions: HCO_3', CO_3''; as well as OH' and H^+ from water. *There are more strong cations than strong anions, and the difference, taking account of valency, is called the alkalinity.* The full definitions of alkalinity contain also minor components such as borates $B(OH)'_4$ and other minor ions but there is no need to go into such detail here. The water in table 3.2 has an alkalinity of 0.00229 mol/kg.

There must always be an equal concentration of anions as cations, so the difference is made up by the weak anions from CO_2. The greater the alkalinity, the greater the capacity for taking up CO_2.

This alkalinity is neutralized by the CO_2 which, with water, produces a mild acid – carbonic acid

$$CO_2 + H_2O \ \rightarrow \ \ HCO_3' + H^+$$

Because of the alkalinity of the sea, sea water holds considerable quantities of CO_2 in the form of bicarbonate ions (HCO_3').

The relationship between CO_2 in the atmosphere and the oceans is governed by a series of chemical equilibria, so there is always a dynamic equilibrium between them. This equilibrium means that as the CO_2 in the atmosphere rises, some of it transfers to the oceans to maintain this equilibrium. In many chemical systems, where equilibria are simple, the relationship is constant, which would mean that when more CO_2 enters the atmosphere, a fixed proportion would be absorbed by the oceans. This, as we will see, is not the case for our seawater system, as the equilibrium is complex, containing a number of reactions.

It is very important to be able to predict the future magnitude of the absorption of CO_2 in the oceans to forecast the magnitude of future global warming. But the problem does not just stop with the absorption capacity of the oceans, because as more CO_2 enters the seas, the acidity (pH) of the seawater changes and this can have effects on the biological systems in the sea which may be catastrophic.

3.2a) The Equilibria involved

Carbon dioxide exists as a molecule in the atmosphere. There is an equilibrium between the amount in the atmosphere and the quantity of CO_2 molecules dissolved in sea water in equilibrium with the air. This is called the gas solubility and is defined by Henry's Law.

Once CO_2 molecules are in water, a reaction with water produces carbonic acid – bicarbonate ions and hydrogen ions (H^+, protons):–

$$CO_2 + H_2O \leftrightarrow HCO_3^{'} + H^+$$

This reaction is reversible and if the solution becomes more acid – the pH is lower – then the bicarbonate ions can revert to CO_2 which can then escape from the liquid, back into the gas phase. The bicarbonate ion can react further with the hydrogen ion to produce carbonate ions – again a reversible reaction

$$HCO_3^{'} + H^+ \leftrightarrow CO_3^{''} + 2H^+$$

The extent of this reaction is determined by the level of hydrogen ion (H^+) in the liquid – ie the pH of the liquid,. The level of the H^+ in aqueous systems is controlled by the ionization equilibrium of water, involving the hydroxyl ion (OH'):

$$H_2O \leftrightarrow H^+ + OH'$$

The level of OH' ion is affected by the alkalinity of the water.

Chapter 3   The Role of the Oceans

The presence of calcium ion and carbonate ions in the water also introduce a further equilibrium whereby these are in equilibrium with unionized $CaCO_3$:–

$$Ca^{++} + CO_3" \leftrightarrow CaCO_3$$

Calcium carbonate ($CaCO_3$) , –the component of limestone, marble, chalk, – is not very soluble in water and so can precipitate out until it remains as a saturated solution. This removes both calcium and carbonate from the system.

$$CaCO_3 \quad \leftrightarrow \quad CaCO_3$$
$$\text{dissolved} \qquad \text{Solid} \downarrow$$

This again is reversible, and solid calcium carbonate can redissolve if sufficient $CaCO_3$ is not present in the surrounding water. To make matters more complicated, there are two forms of calcium carbonate, and they have different solubilities. To further complicate the situation these solubilities are very pressure dependent, meaning that the deep sea will dissolve more $CaCO_3$ than the surface waters. In addition the $CaCO_3$ in solution can be 'supersaturated'. For a number of reasons the concentration of $CaCO_3$ dissolved in seawater is much higher than its solubility. The solution has not been 'seeded' and so the $CaCO_3$ remains in solution – supersaturated – which is very important for the biosystem in the sea.

This is quite a complicated picture. This total carbonate equilibrium system can be expressed mathematically, as all the constants are available, to enable one to predict exactly what happens when CO_3 enters the seawater. The Appendix to this chapter develops these equations and provides a spreadsheet calculation to be able to determine the equilibrium conditions for any atmospheric CO_2 concentration. The relationships have also been expressed diagrammatically in the Bjerrum plot shown as figure 3.2 (Zeebe, 2002).

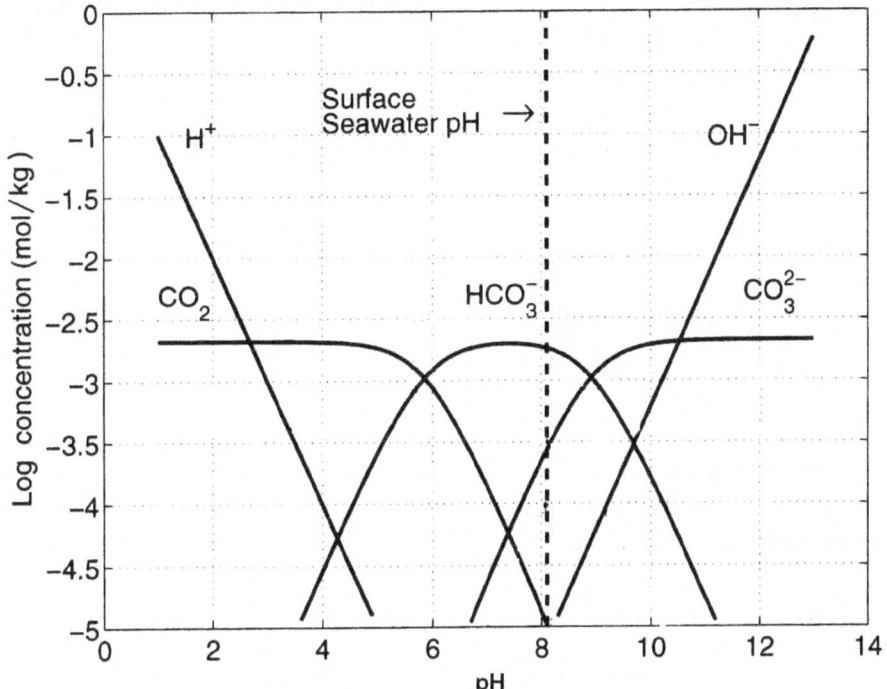

Figure 3.2 The Bjerrum plot – Summarizing sea water carbonate chemistry

The four equilibria associated with the carbonate chemistry of seawater results in concentrations being very dependent on pH. This chart enables the relationships to be understood. In particular, why CO3" is removed from the system at low pH, and the role of the HCO3' ions in controlling the CO3" and the dissolved CO2 concentration. The dissolved CO2 concentration determines the CO2 partial pressure of the seawater.

3.2b) The Revelle Factor

This complex equilibrium picture means that as CO2 is dissolved in the sea it decreases the ability of the sea to take up more CO2 . What we really need to know is the effective absorption, taking this into account. We need to know the relationship between the relative change of CO2 partial pressure and the relative change of the DIC of the seawater. This is sometimes called 'CO2 buffering' and is quantified by the *Revelle factor:*

The Revelle factor (RF) is the relative change in CO2 pressure divided by the relative change in carbon content of the water.

Expressed mathematically:–

$$RF = \frac{d[pCO_2]/[pCO_2]}{d[DIC]/[DIC]}$$

For a normal gas dissolving in a liquid without complex equilibria, the Revelle factor is 1.0. Revelle factors for CO2 and seawater are between 8 and 14, which means that the sea dissolves between one eighth and one fourteenth of the CO2 one might expect.

Using the 4 equilibrium relationships it is possible, very elegantly, to develop an analytical expression for the Revelle factor, so that it can be calculated from equilibrium constants (Zeebe, 2001). The Revelle factor is a concise way of knowing the CO2 absorption capacity of the sea without going through the numerical analysis shown in the Appendix to this chapter.

Since these equilibrium constants have temperature coefficients, the effect of temperature on the Revelle factor can also be calculated. Figure 3.3, shows the resulting relation between RF and CO2 partial pressure, and figure 3.4 gives the temperature dependence.

Using the Revelle factor, we know, when the CO2 in the atmosphere increases, how this affects the sea water carbon content.

Summarizing, we have said earlier that the sea contains 50 times more CO2, than the atmosphere. But if we increase the CO2 level in the atmosphere, we will not get 98% ending up in the sea: a Revelle factor of 10 tells us that only 80% could end up in the sea.

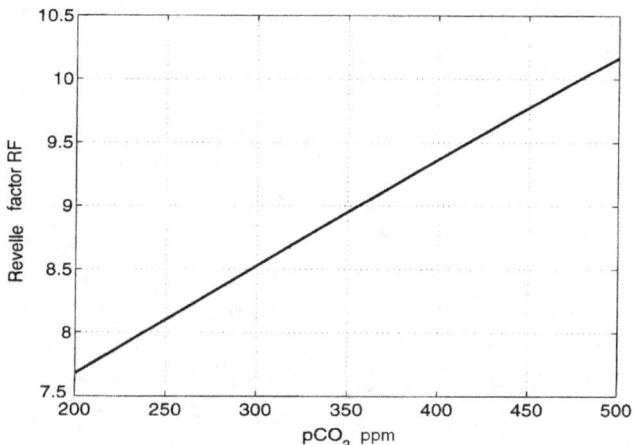

Figure 3.3 The Revelle factor as a Function of CO2 partial pressure
The Revelle factor for a normal simple physical system is 1.0. A higher Revelle factors means that the solubility is that much less than expected because of the complex equilibria.
As the CO_2 partial pressure increases, the increasing Revelle factor means the ability of seawater to absorb CO_2 is decreased.

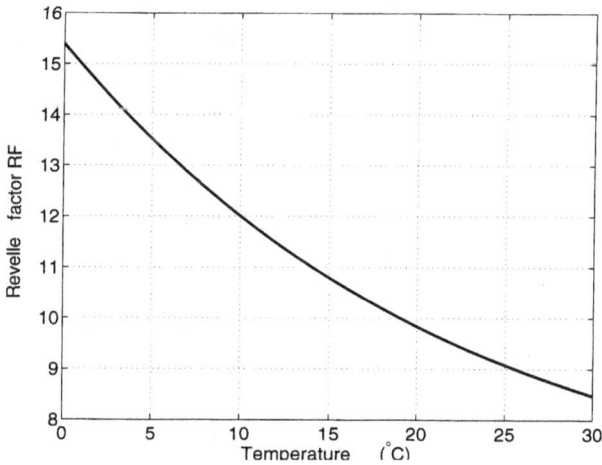

Figure 3.4 The Revelle factor as a function of Temperature
(CO2 360ppmv, Standard seawater composition)
At low temperatures the higher Revelle factor means the ability of seawater to absorb CO2 is less than expected.

3.3) The Biology

The biosystems in the sea have always been most important for the planet. Their life systems involve photosynthesis, and the growth of organisms such as coccolithophorids and foraminifera, which are forms of plankton, the food for all the higher life forms in the sea. The sea floor, in shallow seas, contains a myriad of organisms, including the corals, which are important indicators of a healthy biological system. All these organisms play an important part in the Ca and CO_2 cycles, since they form $CaCO_3$ from their skeletons and form carbonaceous material from their bodies. On the death of these organisms, the $CaCO_3$ and the organic carbon fall to the sea bottom, where much is sequestered and taken out of the cycle. This balances the calcium entering from the rivers and the CO_2 formation from natural events such as volcanoes.

The possibility of global warming gives concern that the biosystem will not thrive at the new temperature levels, but a more important concern is the problems that might occur if the sea composition changes. The previous section explains the chemistry and points out that increasing CO_2 levels decrease the pH and significantly decrease the $CO_3"$ ions in the sea. The effect of this on the biological system may be very serious and is a matter of continuing research.

There are worrying reports on coral reefs turning white, which are considered an indication of poor health. More worrying is the decreased $CO_3"$ level which has been shown to decrease the rate of growth of plankton type organisms ie. those which have a $CaCO_3$ skeleton or shell. As the previous paragraph shows, the higher CO_2 can dissolve the $CaCO_3$ and this would prevent such micro-organisms existing, and so disrupt the whole food chain of the oceans.

This gloomy perspective can be mollified by looking at some of the research on the effects of these changes on the organisms.

Research shows:—
 — that coral reefs have lived successfully at a range of pH values;
 — that white coral is not necessarily unhealthy;

   – that the $CaCO_3$ supersaturation levels are a guard against complete dissolution of $CaCO_3$ based organisms;

   – that the general expected levels of pH change may not be large compared with past variation.

However these are still active areas of research as they may become very important if mankind is determined to burn all the fossil fuel on the planet.

The biological system is active in the 100m of surface waters where light penetrates and photosynthesis is possible, called the *euphotic zone*. The system involves organisms living and dying, resulting in a continuous precipitation of $CaCO_3$ and other organic matter. This material 'rains' down to the sea bottom. In deep waters the pressure increases, increasing the solubility of calcium carbonate so $CaCO_3$ rain dissolves at some level before reaching the bottom. This depth is called the lysocline and the depth is different for calcite and argonite forms of solid $CaCO_3$ because of their different solubilities. So deep ocean beds do not contain calcium carbonate sedimentary rocks as do shallower waters This precipitation of calcium carbonate can be considered as the final step in the carbon cycle which keeps in balance any extra carbon entering the system. Surprisingly, increased CO_2 levels do not increase precipitation but in fact decrease it. (Feely, 2004)

In fact, it is thought that one way in which the planet will finally react to the CO_2 entering from fossil fuels will be for the CO_2 to find its way into the sea, react with solid calcium carbonate, redissolving it as bicarbonate, according to the reactions described in the sections above, and so increase the concentration of Ca^{++} ions and HCO_3^- ions in the oceans. ie the oceans will have a higher alkalinity as the sea bed redissolves.

This process will also occur with the $CaCO_3$ rain now being precipitated, and one likelihood will be a rise in the lysocline level (saturation horizon) where there is no $CaCO_3$ solid. This process may even be important in our lifetime as it is the only known means of sequestering the CO_2, apart from reaction with silicate rocks which take hundreds of thousand of years.

So, within the sea, we have a range of processes which involve CO_2 which alter the composition of the sea itself. There is CO_2 absorption from the atmosphere; CO_2 release to the atmosphere; photosynthesis which takes CO_2 to build organic material; respiration which decomposes organic material to CO_2; calcium carbonate formation and precipitation: and calcium carbonate dissolution as just described. These processes change the amount of inorganic carbon (DIC) in the water and the total alkalinity of the water through Ca^{++} ion changes, all being affected by CO_2 levels and pH. This whole system is elegantly summarized in figure 3.5, taken from Zeebe and Wolf–Gladrow (Zeebe, 2001).

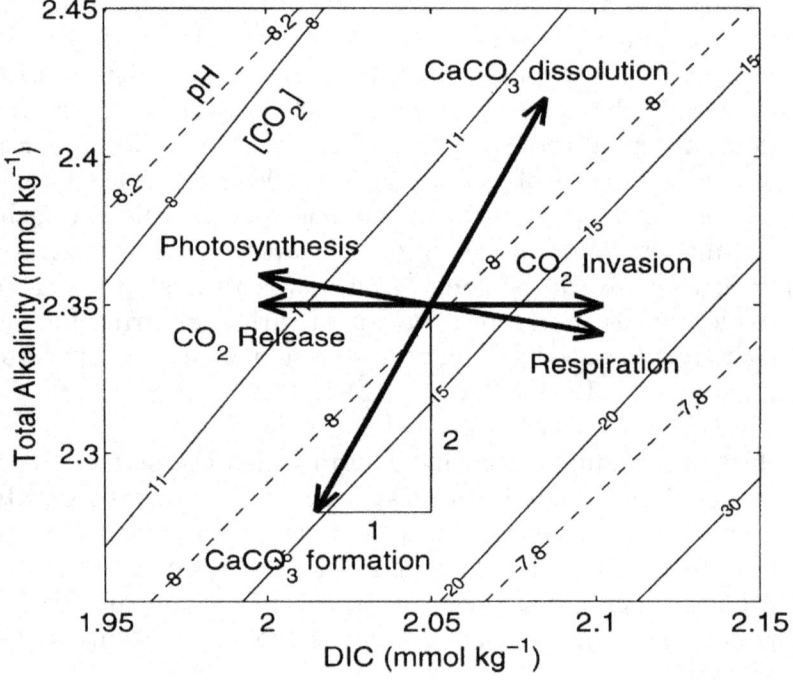

Figure 3.5 Representation of the processes occurring in seawater
The different processes of CO_2 transfer to and from the atmosphere; biological actions of photosynthesis and respiration; $CaCO_3$ precipitation and solution, all disturb the chemical equilibrium of the seawater. This chart summarizes the changes in alkalinity, DIC, and pH when these processes occur.

64

3.4) Mixing

The oceans are vast and very deep with an average depth of 3.6 km and a common depth in the Atlantic of up to 4000m. When discussing the absorption of CO_2 in the oceans, which occur only at the surface, it is necessary to know the mixing pattern of the oceans to determine what capacity there is. Water deep in the oceans can take centuries to mix with the surface waters. The mixing is mainly brought about by ocean currents, which are created by winds and density effects originating from changes in temperature and salinity.

There is a system of ocean currents established on the globe, called the Thermohaline Circulation, which includes the surface Gulf Stream flowing West to East across the Atlantic, and a further surface current going East to West starting in the Pacific, skirting Africa, and crossing the Atlantic where it becomes the Gulf Stream. These surface currents follow global wind patterns, and have been used for centuries by sailors. These surface circulation streams must be part of a closed circulation system and much of the return flows are not surface but deep ocean currents, such as the North Atlantic Deep Water current (NADW). The full cycle is shown on figure 3.6. It is these deep ocean currents that provide the small amount of mixing of deep and surface waters that does occur.

Sunlight can penetrate to a depth of 100 m and it is only up to this depth that biosystems can exist. Below this depth there is no light, and no plant life exists. Down to the first 700 m there is a degree of mixing, caused by surface turbulence, where the waters can be considered to be mixed to some extent, though it may take decades for the mixing to be considered complete.
Below 700 m there is much less mixing, and it can take centuries or even thousands of years to mix with surface waters. Some deeper portions of the oceans are best considered as never contributing to the quantity of seawater as they take thousands of years to mix. But there have been reports of tritium being detected deep in the Atlantic, which can only have arisen from atomic testing done in the Pacific 50 years previously, so there is some mixing in the deep oceans, because of deep currents, but this is a long way off considering there to be good mixing in the technical sense.

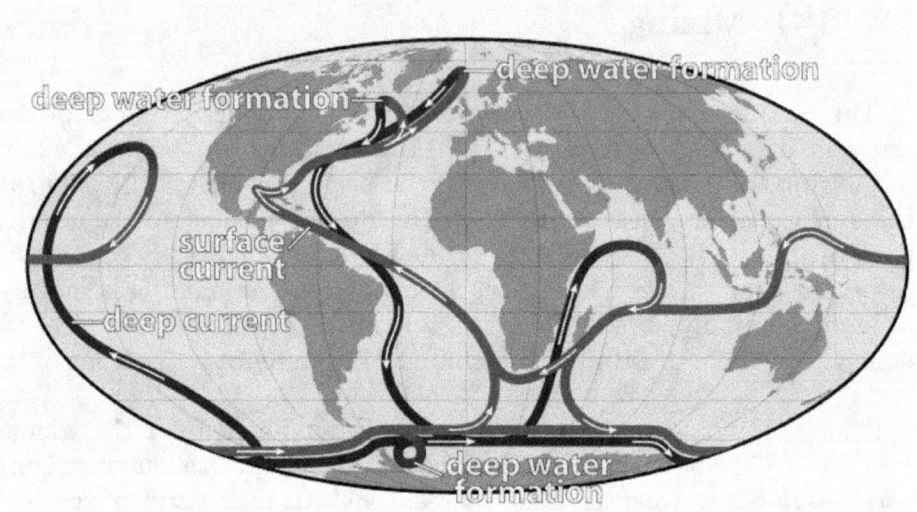

Figure 3.6 Global Thermohaline Circulation pattern
The global surface sea water currents are associated with the trade
sailing routes. The returning waters to maintain circulation are
deep sea currents, and they help with the mixing of deep with
surface waters. Where the deep waters rise to the surface they
create cold currents. Surface waters flowing from equatorial to
cooler regions provide warm water currents – the Gulf Stream for
example.

The most thorough analysis of ocean mixing was carried out in
two international projects, namely the World Ocean Circulation
Experiment, and the Joint Ocean Flux Study. These studies involved
a total of 95 cruises and recordings at nearly 10,000 hydrographic
stations. Results are presented as three separate sections – the
Atlantic, Pacific and Indian Oceans – which together fairly
completely summarize the seas of the planet. Part of these studies
was to determine the fate of the anthropogenic CO_2 which has
entered the oceans since the industrial revolution.   This analysis is
possible using a technique called the ΔC^* technique (Gruber 1996)
using the traces of CFC12 that are dissolved in the sea to determine
the extent of the penetration of surface waters into deeper waters,
and the ΔC is the calculating of preindustrial DIC levels and
subtracting this from the measured DIC.

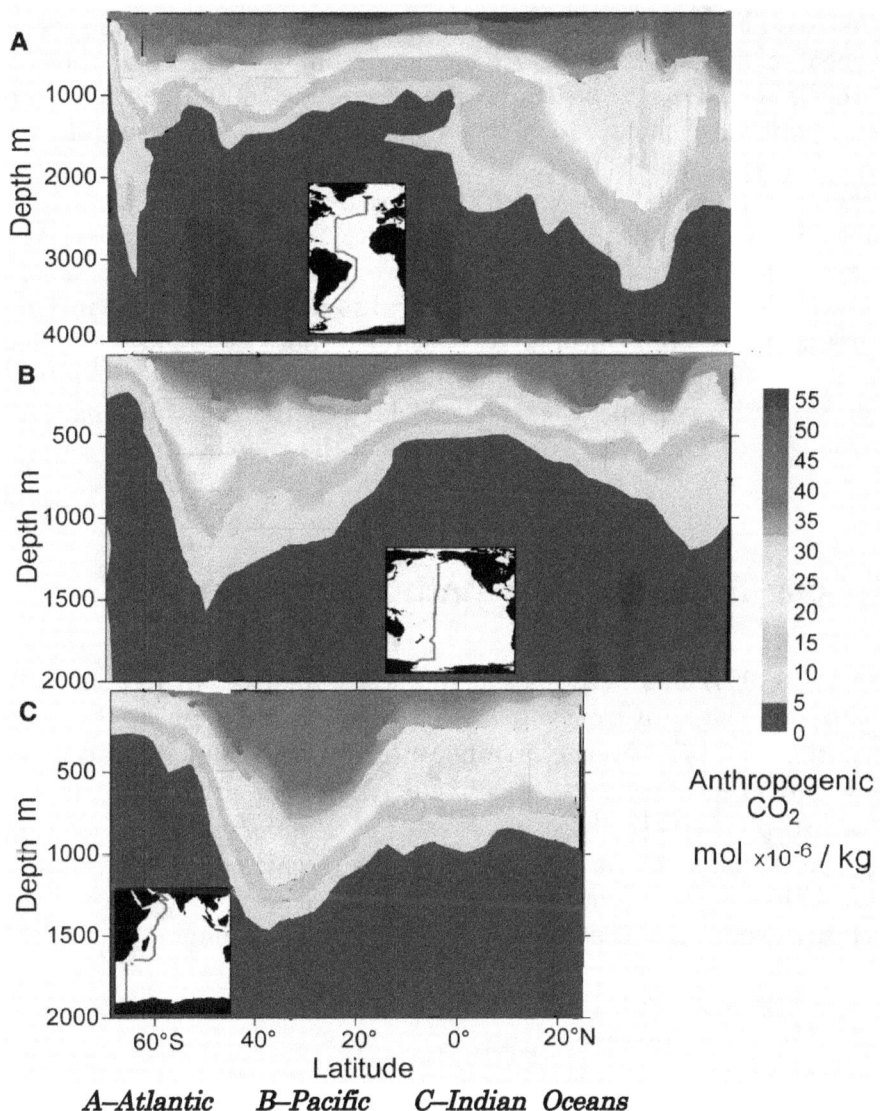

A–Atlantic   B–Pacific   C–Indian Oceans

Figure 3.7 Anthropogenic CO₂ penetration at ocean depths
This figure shows how little mixing there is deep in the oceans; the surfaces are distinctly higher in Anthropogenic CO₂. Deeper than 700m there has been little transfer in the last 50 years, except for fairly deep troughs in all the oceans (the pacific has two). These troughs match with the deep water pattern of the Thermohaline Circulation, shown in figure 3.6.

The result of this work estimated how much CO_2 has been absorbed in the sea and they concluded that 119 ± 19 Giga tons of anthropogenic CO_2 had been absorbed between 1800 and 1994. The observations also show which parts of the oceans are available for absorption and which parts are so stagnant that they have no contribution to make. The results are presented as figure 3.7 which clearly indicate the degree in mixing there has been in the oceans over the last 50 years (Sabine, 2004) . From these figures it looks as though the active depth for CO_2 removal is about 700m, but mixing in this zone is very incomplete because there are much higher concentrations at the surface than at 300 m or deeper. Below 700 m the level anthropogenic CO_2 is very low even after 100 years of anthropogenic CO_2, apart from the paths of the deep ocean currents

3.5) Long Term Sinks for Anthropological CO_2

Archer (1997) has made a specific study of what will eventually happen to the anthropogenic CO_2. There have always been fluctuations of CO_2 in the atmosphere in the past millions of years, but atmospheric CO_2 levels always appear to fall back to the $180 - 280$ ppm level. There appear to be 2 major sinks for CO_2 in geological terms, attained in geological time scales. The first sink is the so–called *CO_2 neutralization reaction* occurring in the oceans which involves the redissolving of precipitated carbonates:–

$$CaCO_3 \quad \rightarrow \quad CaCO_3 + CO_2 + H_2O \quad \rightarrow \quad Ca^{++} + 2HCO_3$$
$$\text{solid} \qquad\qquad \text{dissolved}$$

This leads to an increased alkalinity of the oceans (more dissolved calcium) and a higher DIC (Dissolved Inorganic Carbon) but keeps the pH and CO_2 levels between bounds in which the biosystems survive. This is thought to occur on a time scale of $100 - 300$ years and it is also thought that there are just about sufficient $CaCO_3$ deposits in the oceans to handle the quantities of CO_2 involved.

The second sink is a much slower sequestration reaction by CO_2

68

reacting with silicate rocks, (weathering), where carbonates are produced in place of silicates

$$CaMg(SiO_3)_2 + 2CO_2 \quad \rightarrow \quad CaCO_3 + MgCO_3 + SiO_2$$

This process can occur on land and in the oceans, but its rate is very low, and the time to get back to low CO_2 levels in the atmosphere is of the order of 100,000 years. – this is little consolation for us and no guidance for our politicians.

3.6) Conclusions

The oceans have an important role to play as the final resting place of the anthropogenic CO_2. They have the potential of absorbing the greatest proportion of the CO_2, though it might take decades or centuries to do so. The result of the sea absorbing more CO_2 will make it more acid, and this will change the conditions under which the marine biology has to exist, and its ability to cope is uncertain. There has been some research which shows that the biological systems are fairly robust to pH and temperature, so it may well be an acceptable situation (Pelejero 2005),

More serious is the effect of the CO_2 on the CO_3'' concentrations in the surface waters. CO_2 alters the equilibrium, and produces more bicarbonate and less CO_3''.ions. The marine biology requires carbonate ions for its growth, and so this reduction may well be a serious problem. Furthermore, these biological systems have calcium carbonate skeletons and shells, and they require the calcium carbonate levels in the seawater to be supersaturated to survive. As the degree of supersaturation decreases, there is a danger that the biological systems will not thrive. If $CaCO_3$ is no longer supersaturated, the biological system skeletons will actually dissolve. Since the various forms of plankton are the bread–basket for all the biology of the sea, if CO_2 levels become unfavorable to their growth, there may well be very serious consequences.

The oceans have the potential of solving the fossil CO_2 problem by the so–called CO_2 neutralization reaction, which dissolves precipitated $CaCO_3$, producing 2 mole of HCO_3' and one of Ca^{++} in solution. This will return both the pH and the concentration of CO_3'' in the sea to values before the invasion of anthropogenic CO_2, but with higher level of Ca^{++} ion in the seawater, which is not considered to be a problem for the biosystem. It is generally reported that this reaction is very slow and will take some hundreds of years to complete. This is rather surprising, as it is the same reaction that produces hard water as rain water flows over limestone. It would seem important to investigate more fully the real reaction rate of this important reaction.

Chapter 3 Appendix

Sea Water Carbonate Chemistry

Carbon dioxide exists as a molecule in the atmosphere. There is an equilibrium between the amount in the atmosphere and the quantity of CO_2 molecules dissolved in sea water in equilibrium with the air. This is called the gas solubility and is defined by Henry's Law.

$$CO_{2G} \leftrightarrow CO_{2L}$$

$$[CO_{2L}] = K_H \, p^* \qquad (A3.1)$$

Once CO_2 molecules are in water, a reaction with water produces carbonic acid – bicarbonate ions and hydrogen ions (H^+, protons):–

$$CO_{2L} + H_2O \leftrightarrow HCO_3{}' + H^+$$

$$K_1 = (\,[HCO_3{}']\,[H^+]\,) / ([CO_{2L}]) \qquad (A3.2)$$

This reaction is reversible and if the solution becomes more acid – the pH is lower – then the bicarbonate ions can revert to CO_2 which can then escape from the liquid, back into the gas phase. The bicarbonate ion can react further with the hydrogen ion to produce carbonate ions – again a reversible reaction

$$HCO_3{}' + H^+ \leftrightarrow CO_3{}'' + 2H^+$$

$$K_2 = ([CO_3{}''] \times [H^+]) / ([\,HCO_3{}']) \qquad (A3.3)$$

The extent of this reaction is determined by the level of hydrogen ion (H^+) in the liquid – ie the pH of the liquid,. The level of the H^+ in aqueous systems is controlled by the ionization equilibrium of water, involving the hydroxyl ion (OH'):

$$H_2O \leftrightarrow H^+ + OH'$$

$$Kw = ([H^+] \times [OH']) \qquad (A3.4)$$

The aqueous carbonate system will satisfy all these 4 different equilibria.

In addition, seawater contains more strong anions than cations, the difference between being called the alkalinity. Since the system must be in ionic balance the number of weak cations must be equal to the number of excess strong anions. The total alkalinity (TA) is defined by the difference in strong ionic concentrations, and this must equal the weak cation concentrations.

$$TA = HCO_3' + 2CO_3'' - [H^+] + [OH'] \qquad (A3.5)$$

We have a system of 5 equations and 7 unknowns – HCO_3', CO_3'', H^+, OH', CO_{2G}, CO_{2L}, TA

We must therefore define two variables, and solve to find values for the rest. We can choose which to define, thinking about the use of the model and the ease of solution,. By fixing TA we are defining the seawater we are investigating, and by fixing the pH we are simplifying the solution procedure for the convenience of the spreadsheet calculation. For a water of given alkalinity and pH we can then calculate the carbonate composition. If we want to investigate a particular carbonate composition, then we can rerun the model, altering pH by trial and error to match the condition we are requiring to model.

The Equations

We can define the carbonate ion concentrations from equations A3.2 and A3.3 as

$$[HCO_3' = K_1 ([CO_{2L}]) / [H^+]$$

$$[CO_3^{"}] = K_2 ([HCO_3^{'}]) / [H^+] = K_2 (K_1 [CO_{2L}]) / [H^+}) / [H^+]$$

and the OH' concentrations from equation A3.4

$$[OH'] = K_w / [H^+]$$

we can substitute these values into equation A3.5 to get

$$TA = K_1 ([CO_{2L}])/[H^+] + 2K_2 (K_1 [CO_{2L}]/[H^+])/[H^+]) - [H^+] + K_w/[H^+]$$
$$(A3.6)$$

The only unknown in this equation is $[CO_{2L}]$, so we can rearrange equation A3.6 to solve for it.

$$[CO_{2L}] = (TA + [H^+] - K_w/[H^+])/(K_1 /[H^+] + 2K_2 (K_1/[H^+])/[H^+])$$
$$(A3.7)$$

This set of equations can now be set up in a spreadsheet, giving the location and order of solution. Input data for the spreadsheet are alkalinity and pH, and the remaining 5 variables are calculated within the spreadsheet. To determine the conditions of a particular DIC therefore requires a trial and error calculation, changing the pH until the required DIC is obtained. All the parameter data for the model is available in (Zeebe, 2001).

An example of the output from the model is shown on Table 3.4

This model can be used in a number of ways

- to calculate the CO_2 vapor pressure of any carbonate loading (DIC) seawater
- to calculate the DIC of seawater in equilibrium with any CO_2 atmospheric composition
- to calculate the Revelle factor for any condition (by taking two slightly different DIC cases and computing the changes in CO_2 partial pressure)
- calculate the effect of temperature
- calculate the composition of any aqueous system – for instance rain (TA = 0) is shown to have a pH of 5,4, purely because of the carbonate system.

Verification of the model

Table 3.3 is a copy of the results of the spreadsheet for a number of interesting cases. These show that the trends given by the model are sensible, and by comparing numerical values indicate the accuracy of the numbers. Standard seawater has a pH of 8.1 according to the Bjerrum plot, figure 3.2. Table 3.3 shows a calculated values of 8.14 −8.19. Agreement is quite satisfactory, even though the model described here is much simplified, leaving out minor components such as boron.

Table 3.3
Some Carbonate Chemistry Equilibrium Model Results
Cases taking different atmospheric CO2 concentration in 1900 and 2004

Case	Year	Alkalinity mol/kg	DIC mol/kg	pH	pCO2 ppmv	CO3" mol/kg	HCO3 ' mol/kg
Rain water	1900	0.00E+00	1.12E-05	5.48	280	1.19E-09	3.31E-006
Rain water	2004	0.00E+00	1.45E-05	5.41	375	1.19E-09	3.84E-06
Standard seawater	1900	2.35E-03	2.03E-03	8.19	280	3.15E-04	1.71E-03
Standard seawater	2004	2.39E-03	2.10E-03	8.14	326	2.98E-04	1.79E-03

Table 3.4
An Example of
Seawater Carbonate Equilibrium Spreadsheet Results

line 1 - input data

line2 - Equilibrium constants (Zeebe, 2001)

line3 - results of calculation

temperature	alkalinity	pH	salinity						
2.9830E+02	2.3500E-03	8.1000E+00	3.5000E+01						
-3.0421E+01	-1.3482E+01	-2.0545E+01	-3.5664E+00	6.1445E-14	1.3965E-06	1.1956E-09	2.8285E-02		
2.9830E+02	8.1000E+00	2.3500E-03	7.9433E-09	7.7355E-06	1.0240E-05	3.6204E-04	1.8003E-03	2.7098E-04	2.0815E-03
temp	ph	ta	H+	oh-	co2l	co2g	hco3	co3"	dic

The model, in XML format is available from the author (chemeng@btinternet.com) or from his website

References

Archer, D., Kheshgi, H., Maier–Reimer, E., *Multiple timescales for neutralization of fossil fuel CO2*, Geophysical res, letters, 24, 4, 405-408, Feb 15th 1997

Feely, R. A., *Sabine*, C.L., Lee, K., Berrelson W., Kleypas J., Fabry V. J., Millero, F. J., *Impact of Anthropogenic CO2 on the CaCO3 System in the Oceans,* Science Vol. 305, no. 5682 pp. 362–366, 16 July 2004

Sabine, C.L., Feely, R. A., Gruber, N., Key, Lee, K., R. M., Billister, J. L., Wanninkhof, R., Wong, C. S., Wallace, D. W. R., Tilbrook, B., Millero, F. J., Peng, T., Kozyr, A., Ono, T., Rios, A. F., *The Oceanic Sink for Anthropogenic CO2,* Science Vol. 305, no. 5682 pp. 367– 371, 16 July 2004

Pelejero, C., Calvo, E., McCulloch, M. T., Marshall, J. F., Gagan, M. K., Lough, J. M., Opdyke, B. N., *Preindustrial to Modern Interdecadal Variability in Coral Reef pH*, Science, 309, 5744, 30 Sept, 2005

Gruber, N., J.L. Sarmiento and T.F. Stocker. *An improved method for detecting anthropogenic CO2 in the oceans*, Global Biogeochemical Cycles, 10, 809–837, 1996.

Zeebe R.E., and Wolf–Gladrow D., (2001) CO2 *in seawater:equilibrium, kinetics and isotopes*, Elsevier, Amsterdam

Further Reading

Zeebe R.E., and Wolf–Gladrow D., (2001) CO2 *in seawater:equilibrium, kinetics and isotopes*, Elsevier, Amsterdam

Chapter 4   Observation and Measurement

The increase in CO_2 in the atmosphere began around 1750 – 1800 with the onset of the industrial revolution when coal burning for energy started in earnest. There have therefore been a good 200 years of unnatural CO_2 entering the atmosphere. Hopefully this is time enough to observe and measure any changes it has induced; we need not just rely on looking at the problem theoretically.

The first scientist to consider anthropogenic CO_2 to be a problem, to study it, and also make some measurements was Sten Arhrrenius – in 1898. He was a Swedish scientist who, like many scientists of his day, had a wide range of interests. He was most well known for discovering the relationship between the speed of a chemical reaction and its temperature. which is the famous Arrhenius equation:–

$$k = A e^{-E/RT}$$

His interests ranged to spectroscopy and he was aware of the spectral absorption bands of CO_2. It occurred to him that the CO_2 entering the atmosphere from man's burning of coal would block more infra red radiation (IR) leaving the planet and so cause a temperature rise.

To investigate the effect of more CO_2 he investigated the IR spectrum from the light from the moon both when the moon was vertically above him, and again when the moon was near the horizon. He argued that, when the moon was near the horizon, the moonlight had passed through a thicker layer of CO_2, which was equivalent to increasing the CO_2 concentration. His measurements concluded that doubling the CO_2 would cause a temperature rise of between 5 and 7°C. There is little point in discussing whether this method is really equivalent to doubling the concentration; or that no other factors interfere; or that instruments of the day could measure the differences. What is remarkable is that he devised a method to attempt to measure the effect of increasing CO_2 concentration. It is a most important measurement to make, and it is being attempted now some 100 years later with the help of satellites and very advanced technology.

4.1) Changes in Temperature

The most direct method to determine the influence of increasing CO_2 in the atmosphere is to measure the mean global temperature and see what trend there has been since the onset of the industrial period. Considering the diurnal, seasonal and the geographical variations in temperature, this is no easy task. One organization to respond to this challenge to try to determine the mean global temperature and its trend with time has been a group at the University of East Anglia (UEA) who accept some 8000 readings of temperature around the planet on a regular time schedule and from this develop temperature trends – see Figure 4.1.

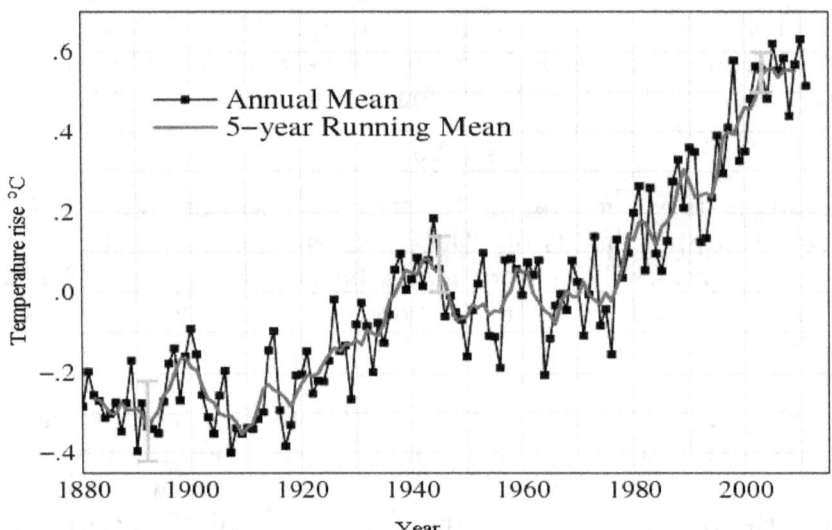

Figure 4.1 Mean planet temperature rise during Industrial Period
There has been a rise over the last 130 years, but because there are so many factors involved, the rise has been erratic, with 2 periods of sharp rise , 1920 – 1940, and 1970 to 2000, the other periods are flat. There has been no significant rise during the first decade of this millennium.

This project requires each weather station to be accredited, with accurate measurement made in a standard way, and all past data must be on the same basis. One particular problem was the change in the type of screen used for temperature measurement since records began, which required a correction of 0.2°C to earlier data to bring the old and new measurements on to the same basis.

A second group has attempted to measure the global temperatures from satellite observations using spectroscopic methods. The agreement between the two completely different methods is excellent – there can be no doubt, that over the last 100 years there has been global warming.

This temperature increase has been by no means steady. For some years there has been little rise, in other years there have been major rises – for instance, considerable rises between 1910 and 1940 and between 1975 and 2000. Between these two periods of temperature increase, the temperatures remained fairly constant. In particular there has been little rise since 2000 to the present 2011.

4.2) Changes in the Atmosphere

4.2a) Trace Components

The atmosphere contains a mixture of different gases. Some of which have always been present naturally, and some have been introduced as waste products by man. These man–introduced are known as anthropological gases. The natural gases nitrogen and oxygen (N_2 and O_2) are transparent to infrared radiation, and therefore have no effect on the heat transfer from the plant surface. CO_2 and water are naturally occurring and these do have absorption spectra in the IR range and therefore are naturally occurring green house gases (GHGs). The remaining components in the atmosphere are at very low concentrations and are termed trace components. About 40 of these are listed by Ramanathan (1998) of which a small number are naturally occurring – ozone, N_2O, methane – but the majority of which have been introduced by man – ie anthropogenic.

Of these 40 trace gases, methane and the chlorofluorocarbons (CFCs) are the most important for our study because they have strong absorption spectra in the IR range. Another characteristic of these important GHGs is that they are stable compounds and will stay in the atmosphere for many years.

Chlorofluorocarbons (CFCs) are man–made chemicals which never occur naturally. They are very stable and are not removed by chemical reaction. Once in the atmosphere they are there for decades. They were developed because of their inert nature and were used in the 1960 –1990 whenever an inert gas or low boiling liquid was needed. This was considered a great improvement on existing materials – for propellants, refrigerants, anesthetics – instead of being sulfur dioxide for refrigeration it could now be CFC22 (CCl_2F_2). In spray cans, as the propellant for producing sprays for polishes, paints and cleaning materials, there was CFC11 (CH_2ClF). A particular CFC, halothane ($CHBrClCF_3$), was developed as a tailor–made compound to be an ideal anesthetic and was the standard anesthetic used in hospitals in the 1970s.

In all these cases, the CFC ended up in the atmosphere, which caused no one any concern because of its inert nature and very low concentration. Eventually they were associated with a depletion of ozone at the South Pole where ozone degradation was catalyzed by CFCs. To overcome this problem of ozone depletion, there was a world conference at Montreal, agreeing the *Montreal Protocol* to phase out the CFCs that damaged the ozone layer and replace them with non–damaging alternatives. Hence production of these CFCs has now been much reduced. Also, the use of CFCs as spray propellants – CFC12 and CFC11 came to a natural end with the discovery that a butane/water mixture serves as a cheap and effective propellant without being inflammable.

The planet does now have a legacy of CFCs in the atmosphere which will remain for decades. Figure 4.2 shows the growth of the CFC concentrations from 1950 to 2010.

Methane is also a GHG because it is a strong absorber of IR, but the source of methane in the atmosphere is not fully understood. It is thought to be partly natural, and partly introduced by man. It occurs naturally from anaerobic (oxygen free) plant decay, and appears as marsh gas. It is also a fossil fuel and vast quantities are contained trapped in the earth as natural gas. At low temperature conditions, it forms hydrates with water and these may be trapped in the Arctic permafrost or in frozen water. These methane hydrates may release methane to the atmosphere when temperatures rise. There may also be continual natural leakage of methane from the natural gas reservoirs. Methane is also produced by flatulence from ruminant animals, which is a colorful source, but not likely to be significant in quantity.

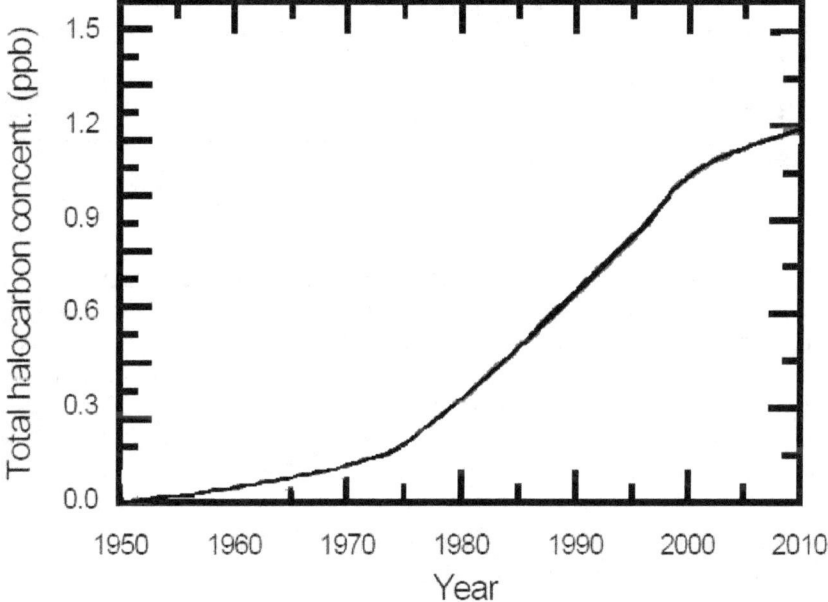

Figure 4.2 The growth of anthropogenic chloro and flouro compounds since the 1950s.
CFCs came into general use during the 1960s. By the 1990 they were recognized as being harmful and withdrawn from production. Their growth in the atmosphere has strongly declined as a result.

Methane release, even from natural sources, could well be caused by mankind as any global temperature rise may release methane locked in various hydrates. Anthropogenic causes include changes in agriculture which result in more methane being produced by animals. The process of collecting and distributing natural gas may well involve leakages and pressure blow–offs. The industry is naturally rather coy on discussing the amount of natural gas escaping into the atmosphere – possible they do not know quantities, but they are making efforts to reduce these 'losses'.

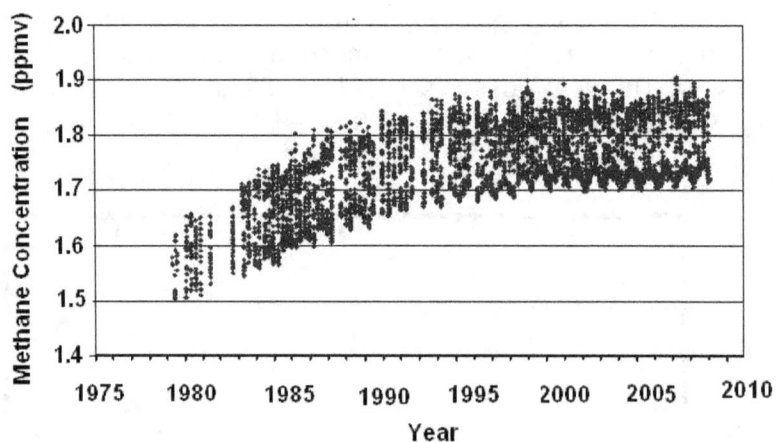

Figure 4.3   Recent Methane content of the atmosphere
There has been steady growth during the last half of the 20th century, but the levels have stabilized during the 1990s and now appear constant. Methane has a short life in the atmosphere, and so the constant composition does not mean the emissions of methane have stopped, but they are now equal to the rate of methane oxidation in the atmosphere.

Methane does oxidize fairly quickly and so has a short life in the atmosphere, but is continually produced from this range of sources so there remains a standing concentration which contributes to the greenhouse forcing.

Table 4.1 shows the concentrations of the various trace components in the atmosphere as of 1994,(IPCC1995) and the radiative forcings they cause – calculated by a radiative model (Myhre 1998). The table shows that CO_2 is responsible for only 60% of the anthropogenic GHG forcings to date. Since the rise in the CFCs has been similar to the rise in CO_2, there are grounds for deniers to claim that CO_2 may not be the cause of our global warming. Lu, (2011) gives a convincing argument from observational data that the warming could be caused by the CFCs rather than CO_2. Because both growth patterns are the same it is not possible to differentiate the relative importance of the CFCs and CO_2 by regression analysis of the observed data, because of statistical correlation. The point should be taken that CO_2 is only part of our global present warming problem – probably little more than half.

Table 4.1 Trace gas atmospheric concentrations and their radiative forcings above their pre–industrial levels

Major Trace Components	Conc. level ppb (IPCC 1995)	Radiative forcing W/m^2	% GHG Forcing
CO_2	358×10^3	1.76	60.48
CH_4	1720	0.62	21.31
N_2O	312	0.15	5.15
CFC–11	0.27	0.09	3.06
CFC–12	0.51	0.2	6.87
CFC–113	0.08	0.03	1.17
CFC–114	0.02	0.01	0.27
CFC–22	0.11	0.02	0.79
CCl_4	0.13	0.02	0.76
CF_4	0.07	0.01	0.24
Total	–	**2.91**	**100**

4.2b) Changes in Trace Gas Concentration CO_2

Carbon Dioxide

CO_2 concentrations in the atmosphere have been measured since the 1950s by the Scripps Institute of Oceanography under the direction of C.D. Keeling, who was the first to seriously investigate atmospheric CO_2 concentration levels. His work shows the steady increasing CO_2 curve, serrated by the seasons as shown by figure 4.4.

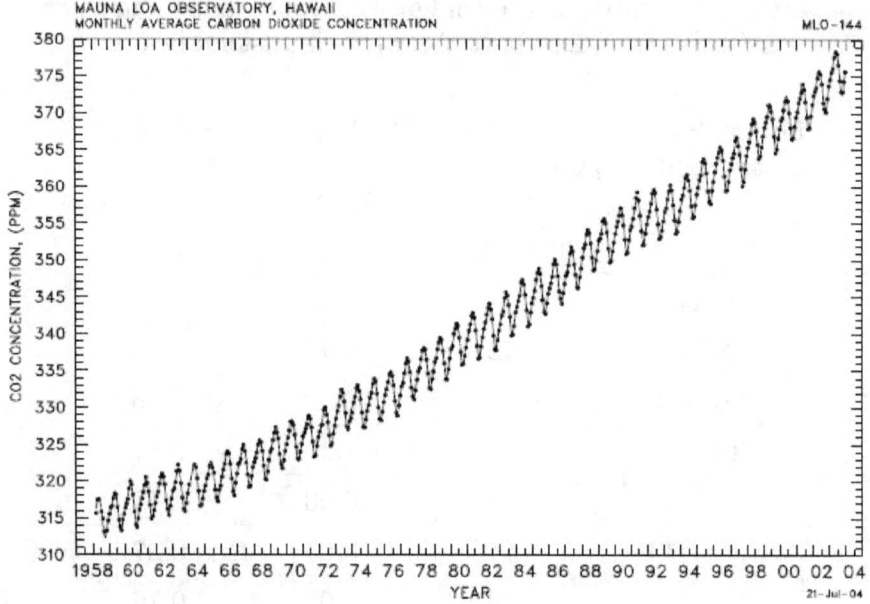

Figure 4.4   The Keeling Curve – CO_2 growth in the atmosphere since 1958

This was the data that alarmed the world of the impending CO_2 concentration growth into new levels for our atmosphere. The serrated pattern reflects the season, where the land bio system grows in the spring, and decays back to CO_2 in the autumn and winter

This work has continued and there is now a very good record of CO_2 concentrations over the last 50 years. CO_2 concentrations in earlier times are measures from air bubbles trapped in ice packs. This gives us a complete record over millions of years. It is clear that since the 1850, there has been a steady rise in CO_2.

Chlorofluorocarbons – CFCs

CFCs were introduced in the 1960s and this can easily be seen in the CFC measured trends as shown in figure 4.2. Over the 1970s the production was expanded rapidly until the 1990s, when legislation required alternatives to be used. This story is reflected in the shape of the curve on figure 4.2. The CFCs have leveled out since 2000 and have even slightly fallen. It appears that the legislation has worked, and the planet will slowly recover from the CFCs as they are depleted by slow reaction over time.

Methane

Methane is difficult. There is no certainty as to where the sources are, or how these sources will be affected by any future changes. The measured trend for methane is shown in figure 4.3

There has been a significant increase in the second half of last century but it appears to have leveled off since 2000. This may be good news, but unless we fully understand where it is coming from, we cannot assume this leveling off will be permanent.

4.2c) IR Radiation Measurements

The spectrum of radiation leaving the planet has been measured by satellites since the 1970s so it is of interest to see how the spectrum has changed, now the atmosphere contains so much more CO_2 and CFCs than it did in 1970. Chen and Harries et al (2007) took measured data from 1970 and 2006 and compared the two spectra, and also compared these spectra with predictions from a theoretical radiative modeling code similar to the MODTRAN4 described in chapter 2.

What spectral differences to be expected are a reduction in the spectral intensity where there are more GHGs, and a raising of the spectral intensity in the two 'spectral windows' because the sum energy loss must remain the same.
There were major difficulties, because the two sets of measurements were made with different instruments, but they satisfactorily

concluded that there were the small changes to be expected, within the spectral signatures in the regions of absorption of CO_2, and CH_4 being reduced, and the intensity in the Windows increasing slightly.

The difference between the two spectra was compared with the difference between the two spectra derived from the radiative model, with the model using the atmospheric compositions of corresponding to the two years 1970 and 2006. The agreement found was very good, as figure 4.5 shows.

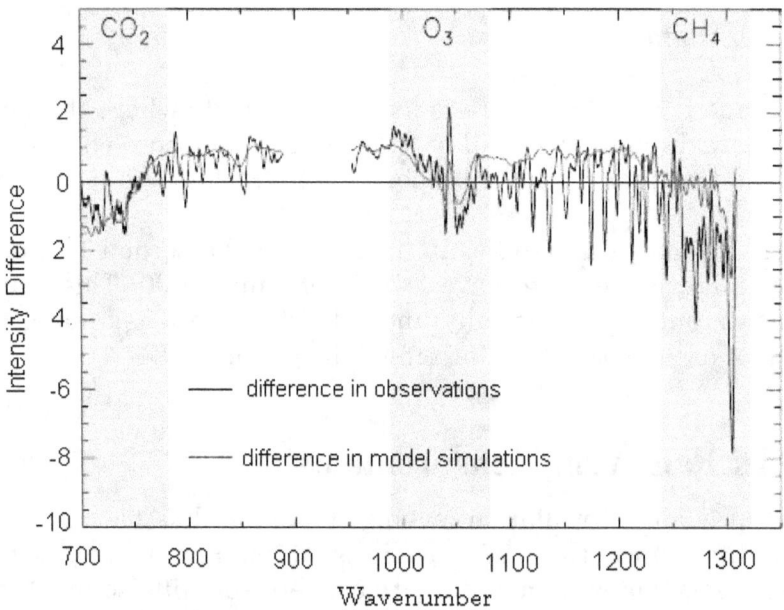

Figure 4.5   The difference TOA IR spectrum between satellite observations made in 1970 and 2006.

The difference in the concentration of trace components in the atmosphere would be expected to show up in observed differences in those parts of the spectrum associated with the trace components. There are small differences to be seen in the regions of the CO_2 and methane spectrum, which could be due to their increased concentration since 1970.

The radiative model predictions for the spectrum differences between the 1970 and 2006 atmospheric concentrations also show the change in spectrum associated with the trace components. The deep methane dip is considered to be observational error.

This work shows that there have been measurable changes to the spectra leaving the planet (TOA spectra) over the last 36 years, and that the predictions of the radiative model agrees well with measurement. Spectral analysis is a highly developed science and its model results can be accepted with confidence.

In a second analysis of TOA measured spectra, Lindzen and Chou (2009) compared this spectra over the pacific ocean over time periods when the sea temperature was rising or falling steadily over a number of months when temperature changes over 0.2°C were observed. There is a strong theoretical argument that as the sea temperature rises the water vapor increases; this absorbs more IR and this leads to a reduced TOA IR spectrum. This effect is the basis of the positive feed back that is predicted by most climate models.

Analysis of reported TOA radiation measurements showed that there was no strong correlation between sea temperature and the TOA spectrum. As will be seen in later chapters, the presence of positive feed backs is a very important premiss on determining the seriousness of future global warming. It is very important to get experimental confirmation of feedback and Lindzen is one of the few who have worked on this aspect; and he found none.

4.3) Changes in the Oceans

4.3a) Sea Level Measurements

There is concern that global warming will cause changes in sea level which may be disastrous. Measurements of sea water level changes to date can give some evidence of the seriousness of the problem. Trends in mean sea water level are not easy to collate from tide gauge records, although attempts have been made and data reported. Over the last 20 years the levels have also been measured by satellite which is thought to be more accurate. Figure 4.6 shows all these measurements. The curve shows there has been clear increases in sea level, but nothing alarming.

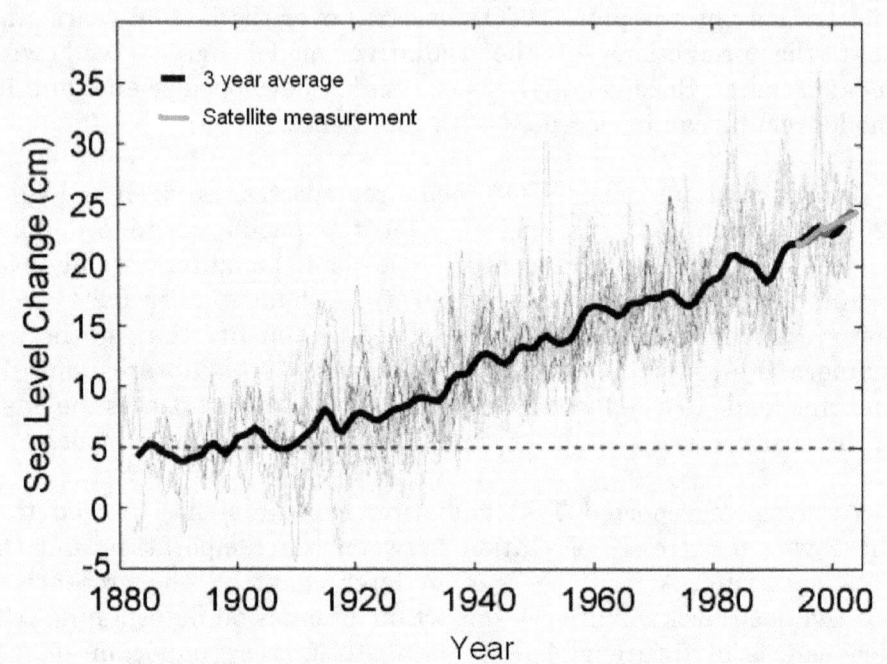

Figure 4.6   Recent Sea Level Rise

A very difficult measurement to take, particularly in the early years. However the 3 year mean does show a steady rise.   The satellite measurements since 1990 agree in slope with the mean of the tide gauge readings.

4.3b) Changes in Polar Ice

Arctic Sea Ice

Over the last 50 years there has been a reduction in the amount of sea ice in the Arctic. Both in the area covered and in its thickness. Data on the thinning of the ice came from submarine patrols under the Arctic who observed that, over the years, the ice thinned from 5 m thick to just 2 or 3 meters. The amount of sea ice fluctuates annually with the seasons, so it is always melting and forming. Over the last 30 years, this procedure has been observed by satellite, and the sea ice maximum recorded yearly. Figure 4.7 shows the satellite images of the maximum extent in the years 1980 and 2007. There has been a steady decrease of about 30% in area over this time, – a mean annual decline of about 1%.

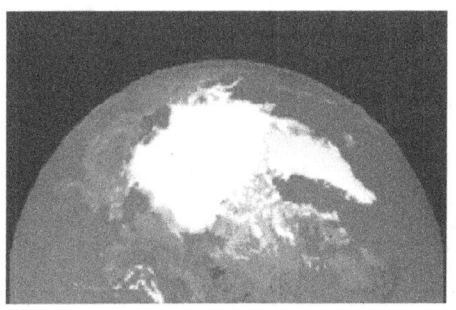 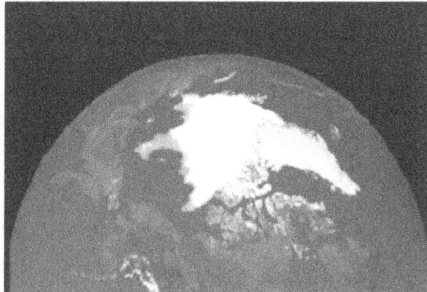

Figure 4.7 The shrinking of the Arctic ice cap
These 2 images of the Arctic were taken at the same time of year but 27 years apart. They show clearly that the arctic sea ice coverage has shrunk, – by about 30%

Glaciation

It is general knowledge that all glaciers on the planet are retreating. Since 1970, there has been a project to record the condition of the world's glaciers, and this confirms that their retreats are significant. In fact there were major retreats in the 1940s 1950s, and since 1980s . There have been stable or slowly growing periods in the 1920s and 1970s. These correspond approximately to the planet temperatures. Figure 4.8 shows the cumulative reduction in mean glacier thickness since 1960. The mean loss in thickness has now reached 13 metres.

Retreating glaciers are a very poignant indication of global warming. I visited a glacier in Switzerland in 1970 and the same glacier in 2000. The retreat is quite impressive. Whereas the Rhone Glacier used to reach the road, it is now about 1 km up the mountain. This retreat can even be looked at quantitatively. The extent of the glacier is a long term summation of the mean temperature at that level. The

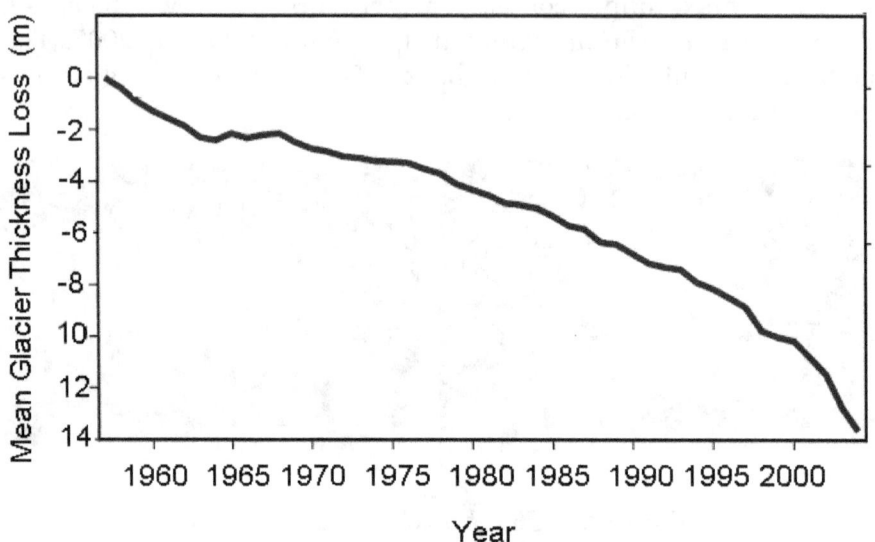

Figure 4.8   Trends in Glacier Thickness
Since the formation of the glacier watch project, considerable data is now available to show how much the glaciers on the planet have shrunk. This chart shows that there has been a reduction almost every year since 1960, and the glaciers are on average 14 m thinner than they were then.

$0.7°C$ global warming that has occurred over this time – see figure 4.1 – will increase the temperature at the original foot of the glacier. To return to the original temperature, the glacier retreats until the temperature has reduced by $0.7°C$. Looking at the adiabatic lapse rate of $6°C$ per 1000m vertical height change, then we are looking at a height change of 120m.   If the glacier is on a slope of one in 10, this means the glacier will retreat 1200m to return to the original temperature conditions at the foot of the glacier. So retreats of this magnitude are to be expected, with the data and measurements that have been recorded.

4.3c) pH levels in the Oceans

The previous chapter discusses the effect of CO_2 on the oceans, and shows that an immediate effect of more CO_2 is to decrease the pH on the seas, which in turn lowers the CO_3'' concentrations, and

changes the supersaturation of $CaCO_3$ – all of which are vital to the seas biosystem. It is of importance to know whether the anthropogenic CO_2 over the last 150 years has in fact changed the mean pH of the seas. This cannot be known from past direct measurements, as pH has not been measured until recently, and the idea of a mean sea pH is as difficult to determine as a mean global temperature, because the pH does vary so much in different seas. Figure 4.9 summarizes the present–day pH range of the world's oceans.

There are reports that sea water pH can have values ranging from 7.5 to 8.4. The pH depends not only on the CO_2 levels but also the alkalinity, which can be altered by differing amounts of Ca^{++}, that enters the sea in river water or by changes in $CaCO_3$ precipitation.

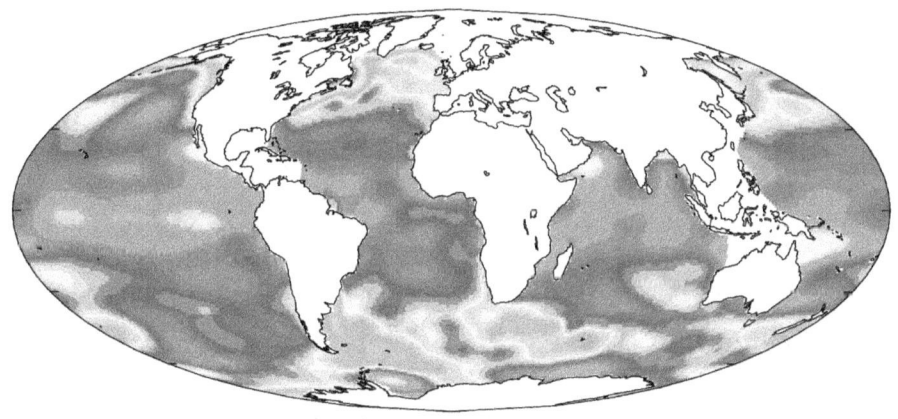

Present day sea–surface pH

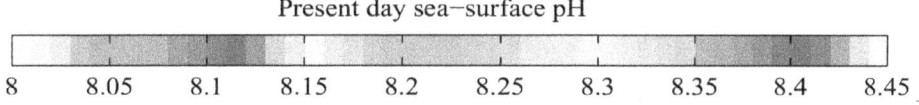

| 8 | 8.05 | 8.1 | 8.15 | 8.2 | 8.25 | 8.3 | 8.35 | 8.4 | 8.45 |

Figure 4.9 *Variation in pH in the oceans of the world*
This gives an understanding of the range of pH which can be considered acceptable for the oceans' bio–systems, which will be useful in discussing the effect of oceans acidity caused by increasing anthropogenic CO_2

One possible method to investigate past pH levels is to look at data from ice borings and use a 'proxy' to see what the past pH was for that one location. A useful 'proxy' for sea pH is the boron isotope [11]B. Chapter 8 explains in some detail how the [11]B isotope can be used to identify the pH of the sea at the time the sea bed was formed. Such research has been carried out with borings on coral reef (Pelejero, 2005) and the results have

shown that over a 300 year period the pH of the pacific reef investigated has had large variations, between 7.9 and 8.2, with a 50 year cycle – see figure 4.10, with no trend to acidity over the last 100 years. The reasons for this fluctuation are possibly due to changes in ocean currents, which also have a 50 year cycle.

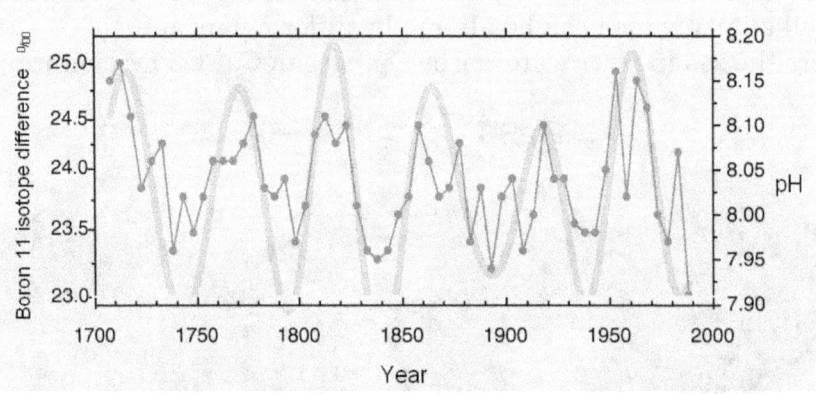

Figure 4.10 300 years of pH variation taken from Pacific Coral Reef Borings.

Using [11]B isotope analysis as a proxy for pH, the pH can be seen to have cycled between 7.9 and 8.2 over the 300 years period on a 50 years cycle. There has been no observable steady decrease as a result of the increased CO_2 in the atmosphere and its absorption in the sea. It is interesting to note that the coral has survived such wide swings in pH, suggesting that coral is hardier than some believe.

The conclusion of this investigation is that coral reefs can thrive with pH variations from 7.9 to 8.2, and there has been no measurable increase in pH over the last 150 years.

There are numerous claims made in the literature that the pH of the sea has already changed by 0.1 units as a result of global warming. It has been calculated, using chemical equilibria, that the pH will have changed by the 0.1 units because the CO_2 levels have changed from 280 t0 360 ppm over the last 150 years – but this is only so if there has been no change in alkalinity. There has been no direct measurement of this 0.1 change in pH (Jacobson, 2005). The actual change depends on the chemistry, the alkalinity, the amount of mixing, and reaction by the sea biosystems. The work of Pelejero is important in that it does not support the pessimistic scenarios that have been suggested will occur with increased CO_2 and global warming.

4.3d) Anthropogenic CO_2 in the Oceans

As described in chapter 3, there has been a number of extensive projects to measure the amount of CO_2 which has transferred into the oceans from the atmosphere in the last 150 years. This work has concluded that between 1800 and 1994, 118 ±19 giga tons of anthropogenic carbon has entered the oceans. Using best estimates of anthropogenic C created over this time period and subtracting the amount remaining in the atmosphere and oceans leaves 39 ±28 giga tons carbon that the authors claim must have been absorbed terrestrially.

These numbers will be discussed in more detail in later chapters. The important conclusion from their work is that the oceans have absorbed about 48% of the CO_2 emitted by fossil fuel, and this has been determined by actual measurement.

4.4) Changes in Climate

Frequency of Extreme Events

There is a general expectation that there will be an increase in extreme weather events as a consequence of global warming. It is fashionable to blame any extreme event on CO_2. Research projects have measured climate variations and analyzed them statistically to determine to what extent there has been an increase in extreme events.

One study has been made by investigating the weather records for many stations in the northern hemisphere, and from this data extract precipitation information to investigate whether there was evidence for an increase in heavy precipitation from observed data (Min, 2011). The mean precipitation index was then shown over the 50 year period. See Fig 4.11 . There has been a steady rise in this Index over this time. The rise is in line with the increase in the vapor pressure of water that can be expected from the global warming that has been measured during this time (about 6% more precipitation from the 1° C rise).

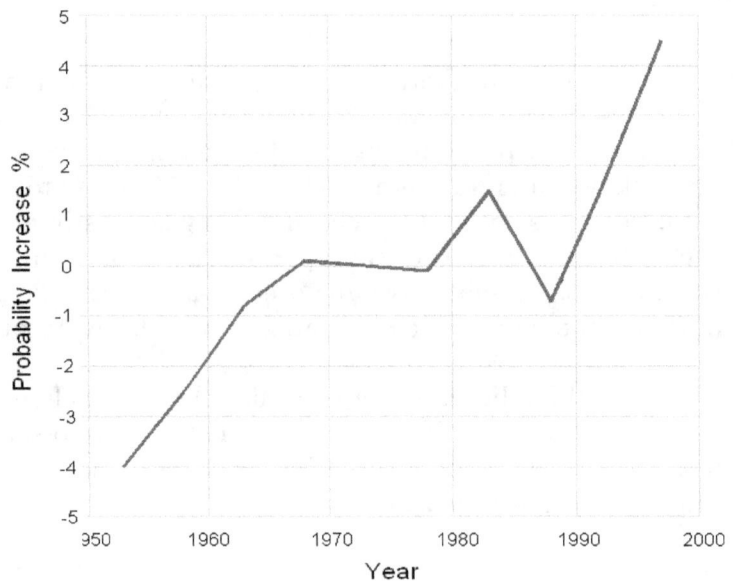

Figure 4.11 Increase in probability of extreme rainfall since 1950
There has been an 8% change in the likelihood of extreme precipitation over the last 50 years. This increase is in line with the 6% higher water vapor content of the atmosphere because of the approximately 1° C global warming that has occurred during this time.

In a number of other published studies, the effect of Sea Surface Temperature (SST) on the likelihood and magnitude of cyclones, hurricanes and high winds in the Atlantic have been studied over the last 40 year period. This is because of the perception of the increase in number and severity of disasters occurring in the Caribbean and the East coast of the US. The last 40 years represent a significant period of global warming, and it is a period where reliable data, particularly SST, data has been collected. These studies could find no statistical relationship between the Sea Surface Temperature and the intensity of events experienced.

A detailed analysis of weather records therefore show that extremes of rainfall can be influenced by global warming, but not by a large amount. There is no evidence that the major changes in weather patterns such as a doubling of the rainfall or very long heat waves can be scientifically or statistically attributed to global warming by CO_2.

4.5) Attributable but Unprovable Effects

All the discussion in this chapter has so far been on observations of events which could be scientifically attributed to global warming. It is very common to attribute any unexpected change to global warming, without seeing any need to provide evidence or argument for the relationship. The world is seeing many more tragedies – monsoons, flooding, hurricanes, tsunamis. droughts, earthquakes – all of which are detailed on the Media, with hourly updates, and frequently blamed on global warming. What has definitely changed is the immediacy of the Media. It is perfectly correct to say that these events are more in our consciousness than they have ever been before, but whether there are many more than previously, or they are just better reported is the question. Looked at with a scientific eye, one should only attribute these events to global warming if there is a scientifically arguable connection. In particular, the one degree rise in planet temperature which has happened, from 288K to 289K is difficult to name as a cause of major catastrophes. One can expect a 6% increase in rainfall because of water vapor pressure effects, but major changes in flow patterns and intensities are unlikely to have been caused by mere 1 °C temperature change.

One particularly unfortunate outcome of blaming all catastrophes on global warming is the tendency for the poorer countries – which seem to have more than their fair share of disasters – to blame it all on the global warming, caused by the rich countries living their energy–rich comfortable life styles. This is deeply engrained in their logic, and will be difficult to counter. Relations between rich and poor will only get worse as we all fight over a decreasing energy pool.

Biologists are not blameless. There is a strong tendency to blame any change in species on global warming. In many cases biological populations fluctuate from a predator/prey relationship; in other cases, small temperature changes can cause migration of species, as they may be dependent on exact conditions. It is unprovable, but often stated that we are in a time of immense decline in the variety of species on the planet. and this must be due to global warming. These conclusions are difficult to justify, since 99% of all species that have occupied the planet are now extinct, and there have been multiple periods of mass–extinction (defined as being when 75% of existing species become extinct) during the history of the planet. We are also discovering more species than we knew existed, so it is difficult to claim we know the present rate of extinction.   It is not very disciplined just to lay all changes at the door of global warming.

Polar bears and collapsing ice cliffs make popular illustrations of global warming. The ice crashing infers the loss of the icecaps. and the wandering polar bears, marooned on pieces of ice denote the reduction in their living space. Both are just cheap publicity. Snow continually falls on Greenland and the Antarctic, and the result is that the ice cliffs must always be falling into the sea to maintain the water balance with the oceans – there is no need to discuss global warming. The polar bears meet a 50% seasonal reduction in their habitat every year. They will be marooned annually on patches of ice, whether or not there is global warming. '*Don't worry about the polar bears – they are doing fine*' is a message from a polar scientist commenting on the doubling in numbers of polar bears in the last 50 years.

Some cause and effects have acceptable logic, but even these can be manipulated by the Media. For, instance, the prevalence of a virus which deforms fetuses of unborn animals (Shellenberg's disease) has been laid at the door of global warming. The virus is spread by mosquito; the 1°C global warming means that the mosquito has spread northwards from southern Europe, hence the disease has now reached mid Europe. So it is correct that mid Europe now has a problem because of global temperature rise which it otherwise would not have had.

So the Media announces – *Global warming causes death and deformed births in our farming stock.*

4.6) Conclusions

This chapter shows that measurements have confirmed that temperatures on the globe are rising; that the weather is becoming more intense though not catastrophically so; that sea level rises are actually measurable; and the atmospheric composition changes with time.

What has still to be discussed is: what is causing this warming? And how grave are the changes likely to be if we do, or do not make any changes to the managing of the planet?

The next chapter tries to investigate quantitatively the importance of the various factors, and looks into the predictions for the future.

References

Chen, C., Harries, J., Brindley, H., Ringer, M., *Spectral signatures of Climate Change in the Earth's' infrared spectrum between 1970 and 2006,* Joint 2007 EUMETSAT Meteorological Satellite Conference and the 15th Satellite Meteorology Oceanography Conference of the American Meteorological Society (2007)

Gruber, N., J.L. Sarmiento and T.F. Stocker. *An improved method for detecting anthropogenic CO2 in the oceans*, Global Biogeochemical Cycles, 10, 809–837, 1996.

IPCC, Working Group I, to the Second Assessment Report of the Intergovernmental Panel on Climate Change, *The Science of Climate Change,* 1995

Jacobson, M. Z.,*Studying ocean acidification with conservative, stable numerical schemes for non–equilibrium air–ocean exchange and ocean equilibrium chemistry,*Journal of geophysical research, *Vol. 110, D07302, 17 PP., 2005*

Lindzen, R. S., Choi, Y., *On the determination of climate feedbacks from ERBE data* , Geophys . Res. Lett., 36, L16705, 2009

Lu Q, (2010), *What is the major Culprit for Global Warming CFCs or CO2?,* J of Cosmology, 8, 1845–1862

Min, S., Zhang, X., Zwiers, F.W., and Hegert, G. C., (2011), Human contribution to more intense precipitation extremes, Nature, 470, 378–381

Myhre, G., Highwood, J. H. Shine, K.,and Stordal, F., (1998), *New estimates of radiative forcing due to well mixed greenhouse gases,* Geophysical Res. Letters, 25, 2715–2718

Pelejero, C., Calvo, E., McCulloch, M. T., Marshall, J. F., Gagan, M. K., Lough, J. M., Opdyke, B. N.,   *Preindustrial to Modern Interdecadal Variability in Coral Reef pH*, Science, 309, 5744, 30 Sept, 2005

Ramanathan, V., Trade–Gas greenhouse effect and Global Warming, Ambio, 27, 3 May , 187–197, 1998

Sabine, C.L., Feely, R. A., Gruber, N., Key, Lee, K., R. M., Billister, J. L., Wanninkhof, R., Wong, C. S., Wallace, D. W. R., Tilbrook, B., Millero, F. J., Peng, T., Kozyr, A., Ono, T., Rios, A. F.,*The Oceanic Sink for Anthropogenic CO2,* Science Vol. 305, no. 5682 pp. 367–371, 16 July 2004

Chapter 5   Explaining the Past and Predicting the Future

5.1) Mathematical Modeling Principles

There has been a problem of increasing CO_2 and global warming for more than 100 years. Measurements have been taken over this time and so there is ample quantitative information assembled. Natural laws and theories regarding mass transfer, chemical equilibrium and radiation abound, and so it is a matter of finding out whether we understand what is happening by comparing quantitatively the outcomes of the last century with what we would expect from theory. If we can properly match calculation with measured data, we can feel confident that we understand the physics of the situation and we can predict future changes with some confidence. If we are unable to match calculations with measured data, this shows we do not fully understand the system and so predictions cannot be taken seriously. The technique for comparing measured data with calculation is called 'mathematical modeling'.

Modeling is a well–used technique in science and engineering. It is good practice in mathematical modeling to start with simple models, but include the correct mathematical descriptions in terms of the physics and chemistry involved. Keeping the model simple means combining effects, with much averaging and 'lumping' of parameters, but still keeping the major physical characteristics without going into much detail . This 'simple model first' approach is preferred because it concentrates on the major aspects, and the results are not too complex to understand. Once it is found that the calculations cannot properly fit the measured data, the discrepancies indicate the refinements to be made to the model to provide a better fit. This stepwise approach is used until a fit as good as the experimental accuracy is found.   This results in an adequate model with a minimum number of adjustable parameters. Once a model becomes complex, then a good fit is inevitable because there are so many parameters available to adjust, and the results are often too difficult to interpret. Little can be learned from the fact that a complex model fits experimental results, unless the complexity has been introduced as the only way of getting the model to fit.

5.2) Developing a Simple Mechanistic Model

To start with a structurally correct model, let us assume we have CO_2 in the atmosphere, transferring into the well–mixed sea in accordance with standard gas/liquids mass transfer theory, with the equilibrium back-pressures determined by the $CO_2/H_2O/CO_3''/HCO_3'$ equilibrium (see Chapter 3 appendix) . We can determine a radiative forcing for the resulting atmospheric composition after mass transfer, which we can convert into a temperature rise and global warming, as described in Chapter 2. We will start with this model and see how well we can explain the measurements that have been collected together and reported in Chapter 4.

5.2a) Modeling the CO_2 Transfer to Oceans

The sea contains 50 times as much CO_2 as the atmosphere, and so it could be expected that the majority of the CO_2 will transfer to the oceans.   But evidence shows much of the CO_2 is remaining in the atmosphere. Early estimates suggested that 60 – 70% stayed in the atmosphere, but later estimates revised this to 60 or even 50% .Various reasons contribute to this unexpectedly low absorption:–

1) The complex chemical equilibrium of seawater means that the capacity for seawater to absorb more CO_2 is reduced by a factor of about 10 – the Revelle factor (Zeebe, 2001) for any given atmospheric partial pressure change. Hence one cannot expect the sea to absorb more than 80% of CO_2 entering the atmosphere.

2) The sea is not well mixed, and below 700 m the mixing takes decades, and mixing in the ocean depths may take centuries. Hence utilizing the capacity of the whole seawater is a very slow process.

3) The mass transfer between the atmosphere and the sea may be very poor. With there being one m^2 of transfer surface area per 100km of air and 4 km of water, the crucial 'specific mass transfer area per volume of system' is one millionth of the value of a well–designed industrial absorber,.

The data we are working with are the annual production of anthropogenic CO_2 discussed in Chapter 1 and the annual increase in CO_2 in the atmosphere, as discussed in Chapter 4. This data is summarized in the graphs, given in the earlier chapters, but repeated here, for completeness, as figures 5.1 and 5.2

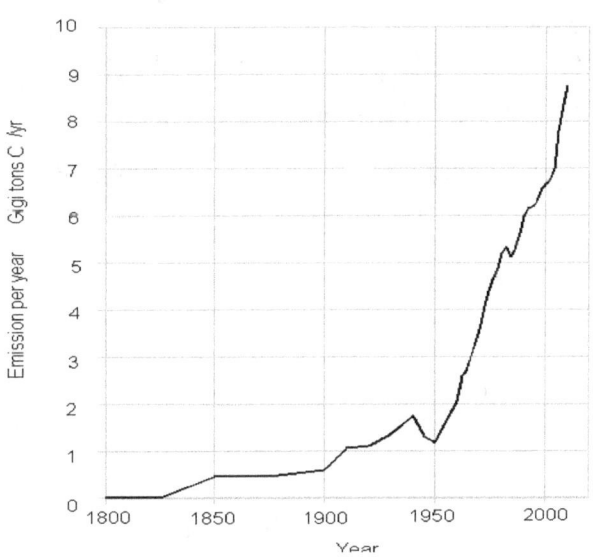

Figure 5.1 Emissions of fossil fuel CO_2 during the industrial period
Data source: These are figures published by CDIAC, Carbon Dioxide Information Analysis Center (CDIAC, 2011)

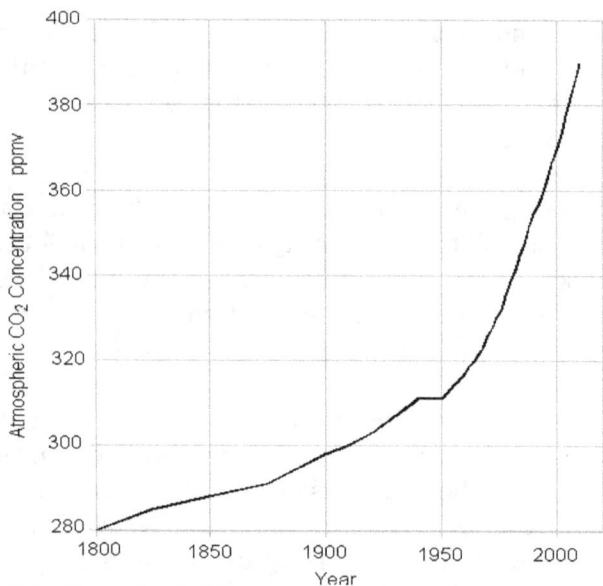

Figure 5.2 Growth of CO2 atmospheric concentration during the industrial period.

Data from the Keeling curve, augmented by estimates of CO2 concentrations before 1958, mainly from ice boring measurements. Data sources: Mauna Loa 2012 for 1850 – 1982, Keeling for 1958 – 2010(Scripps).

The mass transfer model describing the transfer of CO2 between the atmosphere and the sea is developed in the Appendix to this chapter. It assumes that as the CO2 transfers to the sea water, this increases the back CO2 pressure of the seawater which reduce the CO2 transfer that can occur. This whole simple model can be written with only one parameter, the rate at which CO2 can transfer from atmosphere to the sea per sq m of sea surface, which is unknown, and this parameter is altered to find the best fit to the atmospheric CO2 curve.

A value for this one parameter in the model (*Kga*) of $1.53 \ 10^{8}$ (tns CO2/year/ppm) gave the fit shown on Figure 5.3a. In modeling terms, it is quite remarkable that such a good fit can be obtained with such a simple model. It is apparent that the relationship between CO2 in the atmosphere and the sea can be very well described by this simple model. This means we do not need to refine our mass transfer model, to use it with the fitted parameter for predictive work. We can also conclude that, from a statistical point of view, there is nothing to be gained from using a more complex mass transfer model because the results are already predicted within experimental accuracy. A more complex model provides more parameters, and will only confuse predictions, because the model would contain more parameters than justified statistically, given the limited range of data we have available.

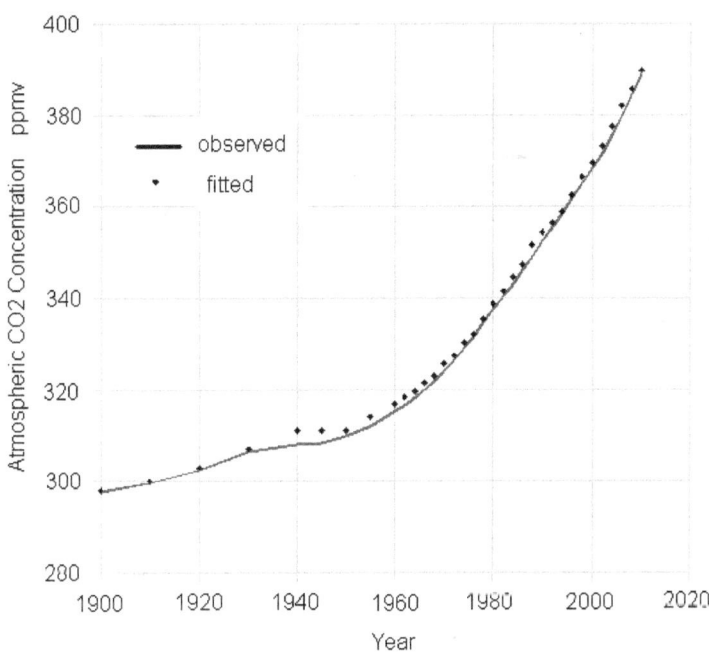

Figure 5.3a Fitted CO_2 atmospheric concentrations allowing for absorption by 4000m depth of ocean.
This remarkably good fit has been obtained using a mass transfer model with only one adjustable parameter. The model clearly accurately calculates the amount of CO_2 that transfers out of the atmosphere, presumably mainly into the oceans. But this model assumes that the whole of the 4000m of ocean depth is active in absorbing the CO_2, which is not considered likely.

In calculating the atmospheric CO_2 curve, the model has to calculate the CO_2 absorber by the sea. This model calculates that 124 Giga tons were absorbed between 1800 and 1994. This compares well with the estimate from measurement of 10,000 samples and 95 cruises which concluded that during this time 119±19 giga tons of anthropogenic CO_2 has been absorbed by the sea — see Sabine Chapter 3.

So we have a semi–theoretical fitted model which can be used for predictive work!

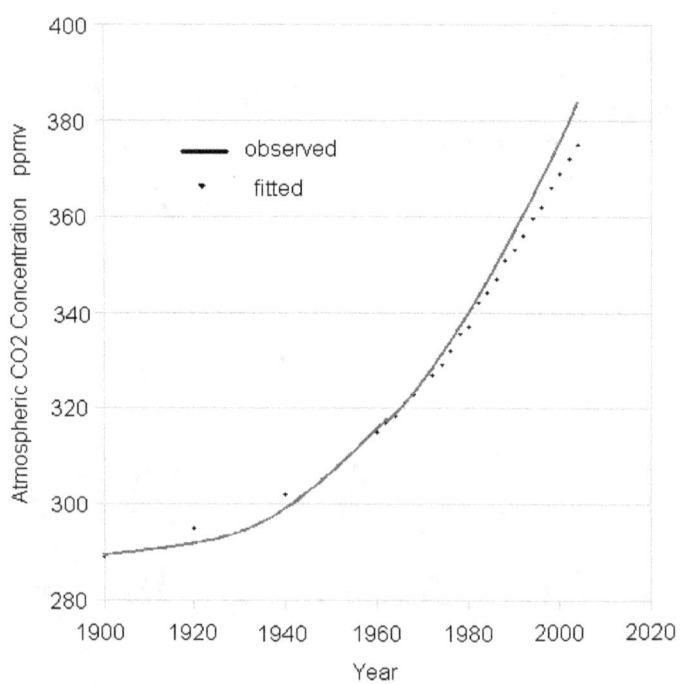

Figure 5.3a Fitted CO₂ atmospheric concentrations allowing for absorption by 4000m depth of ocean.

This remarkably good fit has been obtained using a mass transfer model with only one adjustable parameter. The model clearly accurately calculates the amount of CO₂ that transfers out of the atmosphere, presumably mainly into the oceans. But this model assumes that the whole of the 4000m of ocean depth is active in absorbing the CO₂, which is not considered likely.

The model has assumed that the whole of the 4000 m depth of sea water is available to distribute the absorber CO_2. But we know from the work of Sabine et al (2004) that the anthropogenic CO_2 does not venture below 700m, and the rest of the seawater is best considered as not being part of the active seawater system because mixing takes thousands of years. If we repeat the model fitting with 700 m in place of 4000m, which is reducing the seawater volume by almost a factor of 6, then the best possible fit we can get by altering K_ga is as shown on figure 5.3b. There is no way of getting a better fit with this model. The absorption by the sea is better than the model can predict at higher CO_2 levels.

We need to look for an explanation for this improved mass transfer. Since it is fairly certain that the majority of the seawater is not involved in diluting the absorbed CO_2, it can only mean that the surface waters are able to absorb more CO_2 for a given CO_2 partial pressure. This can only be possible by changing the alkalinity of the seawater. Increasing the Ca^{++} content of the sea will do this, and this will occur if the absorbed CO_2 redissolves the precipitated $CaCO_3$, by the known reaction:

$$CaCO_3 \rightarrow CaCO_3 + CO_2 + H_2O \rightarrow Ca^{++} + 2HCO_3'$$

solid   dissolved   increased alkalinity

The great unknown about this reaction is its speed, which is determined by mass transfer solution of the solid $CaCO_3$ rather than chemical kinetics. Archer (1997) claims that it reacts on a 300 year timescale, but on land the rainwater can become 'hard' (which is the same reaction) within days of exposure to limestone. If we assume that in seawater the reaction only takes years rather than centuries, then this will provide the explanation we are looking for. This is a very significant observation, because if it can be shown to be correct, then this is the most significant factor in mitigating the CO_2 and global warming problems. The evidence presented here is a useful indicator: There needs to be more work done.

This subject is pursued more quantitatively in Appendix II at the end of the book. Since it is no more than a hypothesis, we will not discuss it further in the main body of this text.

5.2b) Calculating the resulting Global Warming

The model calculates the CO_2 content of the atmosphere at any time, and this must be converted into the anthropogenic global warming. We can identify 3 contributions to the temperature rise:

1) the forcing due to the CO_2 level in the atmosphere

2) the forcings due to other GHG in the atmosphere

3) a feedback temperature rise from the increased water vapor content of the atmosphere.

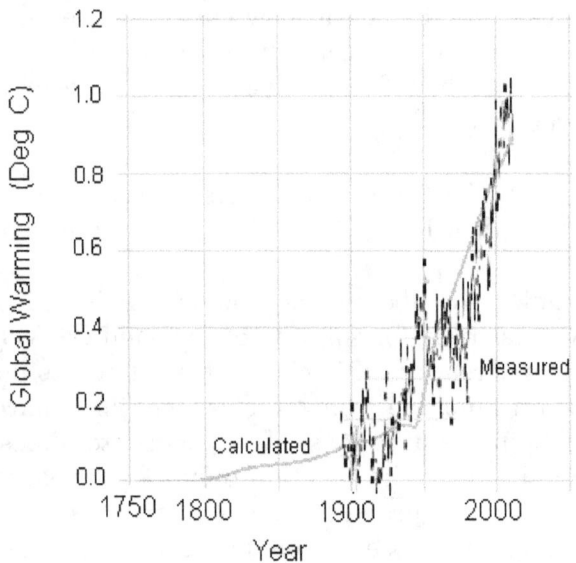

Figure 5.4 Calculated Global warming up to the present using a simple radiative forcing model
The observed data is far from stable because it contains noise from all effects that influence global temperature. The steadier rise from 1970 is fitted by the simple model, which from this date contains a considerable proportion from CFCs and CH4 radiative forcings.

The Chapter 5 Appendix shows how these three contributions can be calculated from the information described in Chapter 2, and added together to provide a predicted global warming from the CO_2 levels calculated by the model. The resulting predictions are compared with the actual measured data in Figure 5.4 This Figure shows the model fits the measured global warming well. The agreement is as close as can be expected, considering the large variations that are present in the measured data.

5.2c) Sea Levels Rise

From the global temperature rise, sea level changes can be calculated, on the assumption that the majority of the sea level change is the result of the density change of water (IPCC,2007).

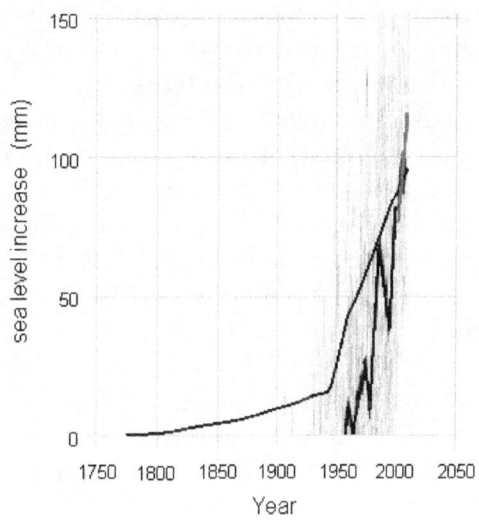

Figure 5.5 Rise in sea level from 1900 to the present day
The figure shows the observed mean sea level rise (thick line), and satellite measurement over the last 20 years, compared to the calculated level changes based solely on density changes due to the global warming, assuming an active sea depth of 700m (continuous thinner line).

The measured sea level appears to rise twice as fast as the model predicts. This may be because more than 700 m depth of the oceans experience a temperature rise, or the thinning of the glaciers and ice caps are contributing to the rise – or both.

Taking an active depth for the sea of 700 m, and the relationship between density and temperature for seawater (Zeebe, 2001), the expansion can be calculated from the temperature rises to give Figure 5.5. The measured seawater level is rising twice as fast as this theory would suggest. Therefore there may be other contributions such as melting ice from the Poles and glaciers.

5.2d) Extreme Events

As observed in the previous chapter, extreme events are affected by the mean global temperature. With an increase in sea surface temperature (SST), the water content of the atmosphere will increase because the vapor pressure of water will be higher, giving a corresponding increase in precipitation. Heat waves will be hotter and floods and storms will deposit more rain. Assuming weather patterns have a random variability corresponding to the Normal Distribution, we can calculate the changes in probabilities of extreme events using Normal tables, and compare 2000 with 1900. Taking a '1chance in 100' (probability 1:100) event for 1900 for heat wave, if we increase the mean temperature by 0.8 K for 2000, this will shift the normal curve to give a '1 chance in 65' (probability 1:65) for that same heat wave definition temperature.

For precipitation we can look at the difference in water partial pressures in 1900 and 2000 and show that a precipitation level with a 1:100 probability in 1900 would have a probability of 1:70 in 2000.

More results are shown in Table 5.1 The extreme precipitation results shown in this table are in general agreement with the reported increases in Precipitation Index that has occurred during the last hundred years in the Northern Hemisphere (Min, 2011). it appears that there has been a doubling of the probability of extreme events occurring. Though this is important for the insurance industry, such a change is difficult to notice without the collection and analysis of measured data.
It is not possible to scientifically justify claims that the wild changes in weather we sometimes experience are due to greenhouse effects.

Table 5.1   Probability of Extreme Events

Year	1900	2000	2100
Event			
Extreme precipitation	1 in 100	1 in 70	1 in 53
Heavy precipitation	1 in 20	1 in 17	1 in 14
Extreme Heatwave	1 in 100	1 in 65	1 in 40
Heatwave	1.in 10	1 in 8	1 in 6

5.3) Predicting the Future

So we now have an adequate model which considers both the mass transfer and the radiative global warming; it can be used to predict future conditions which could apply on the planet after the burning of much more fossil fuel.

In predicting future global warming it is essential to have some idea of the future CO_2 emissions that the planet will have to cope with. Predicting the future is not an exact science, but it is a well developed technique. It is often used by engineers, because all engineering projects must have future predictions available before the capacity of any new engineering facility can be decided. Techniques are built on defining a common sense scenario of the future, followed by the necessary calculations to see what that implies numerically to make the forecast.

In the case of fossil fuel usage, it is reasonable to assume the world will adopt a more or less 'business as usual' scenario, where the developed world (USA, Europe, Middle East), save a little by reducing energy demand by using renewables, inspired by a steady price rise for fuel – so each year a small saving over the previous years consumption might be achieved. The developing world (China, India, Brazil and Africa) will see no reason to hold back demand until it attains the developed world levels – a morally defensible policy – at which point it may make the same savings as the developed world.

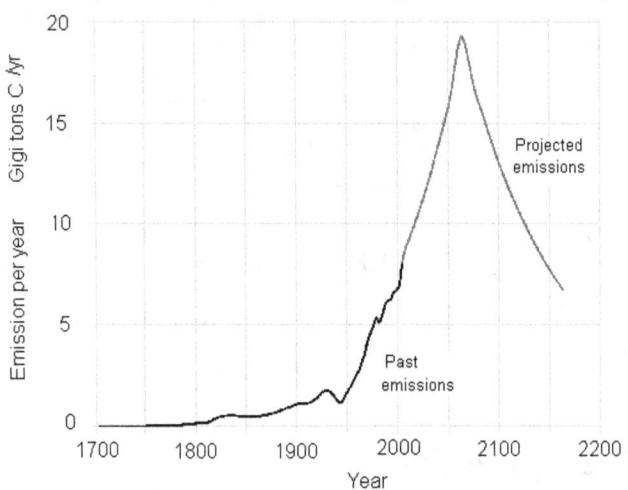

Figure 5.6 A global anthropogenic CO_2 emission scenario for our future global warming study

Future predictions of carbon released from fossil fuel burning (post 2010). Based on assuming a trebling of the developing world consumption at 4% annual increase and a falling of the developed world consumption of 1% per annum; the developing world also falls at 1% per annum after achieving 300% growth. Some predictions suggest economically accessible resources of fossil fuel would be expended by around 2100 using this scenario. The curve is unrealistically sharp, which is the result of the simplistic formula on which it is based, but there are advantages in keeping the concepts simple.

Putting some numbers on this scenario, we could assume a 1% CO_2 saving per year by the developed world – some countries are already forecasting 3% per annum targets, but this is grossly optimistic. The developing world, now with 8% growth rates, may settle down to something less – say 4% per annum until they reach the Western levels of consumption. We can get an estimate of the magnitude of the two worlds by noting that in 2010 the consumption from the developing world was reported as having overtaken the consumption of the developed world – so let us assume of the 8 Giga tons of carbon released in 2009, the developed world and the developing world each released 4 Giga tons of carbon into the atmosphere. The developing world has about 3 times the population of the developed world, so a ceiling for their release would be 12 Giga tns/yr, and they reach this ceiling at the growth rate of 4% per annum. When the numbers are worked out, this scenario results in figure 5.6. It is an understandable and defensible forecast which we will use throughout this book when discussing the future.

This scenario suggests that in the next 150 years fossil fuels will have been largely replaced, but this replacement will be done at an achievable rate with minimal impact on world economic growth. Admittedly, the curve is unrealistically sharp, which is the result of the simplistic formula on which it is based, but there are advantages of understanding and interpretation in keeping the concepts simple.

The International Panel on Climate Change (IPCC) has presented such a wide range of scenarios it is confusing to use them for any working basis (IPCC, 2000). Is the IPCC being so careful to show itself to be a scientific and not a political body, that it dare not stray into the realm of predicting the future? By chance, one IPCC suggested future which compares well with the one presented in figure 5.6 is their 'A1B scenario'.

Taking the future fossil carbon release scenario represented by figure 5.6, our model can be used to give some idea of the resulting future atmospheric CO_2 levels, resulting global warming, sea Dissolved Inorganic Carbon (*DIC*) content, sea acidity, and frequency of extreme events.

5.3a) Future atmospheric CO_2 levels

Running our fitted mass transfer model, with the sea absorbing CO_2 proportional to the CO_2 driving force, produces a CO_2 ppm in the atmosphere as shown by Figure 5.7. This shows the CO_2 reaching a peak of 620ppm, after which it begins to fall, as the release of CO_2 falls below the transfer rate into sea.

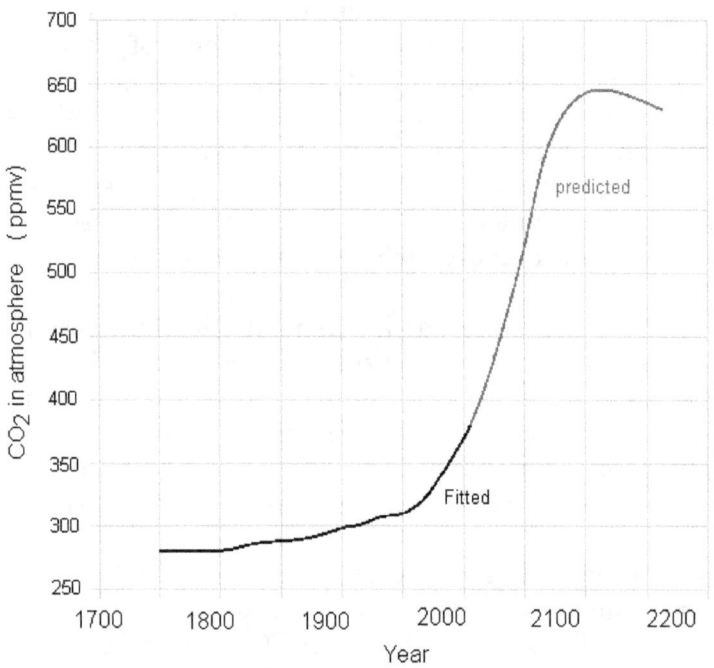

Figure 5.7 Future CO_2 Concentration in the atmosphere, based on predicted CO_2 emissions of Figure 5.6
Using the mass transfer model to predict the CO_2 transfer to the oceans, leaves the quantity of CO_2 shown here remaining in the atmosphere.

This CO_2 peak is surprisingly low, considering the amount of discussion in the literature   mentioning 1000 and 2000 ppm as possible levels. The reason is that mass transfer depends on the driving force. Initially mass transfer is very little because the driving force is very little, but as the atmospheric concentration builds up the mass transfer rate increases significantly. Initially the CO_2 is 'accumulating' in the atmosphere, building up a driving force, which later provides a more significant mass transfer rate into the sea.

5.3b) Future Global Warming

Using the predicted future CO2 atmospheric levels shown in Fig 5.7 it is possible to estimate the radiative forcing involved, and so predict the future global warming, using the method described in the previous section. There is the forcing from CO2, which is concentration–dependent following the decaying Beer's law relationship; there are the CFC and CH4 forcings which we can only assume remain constant in the future; and there is the water vapor contribution as its vapor pressure rises with temperature, which we can include as a feedback. When calculated out in the spreadsheet model described in Chapter 5 Appendix we get a prediction of the expected global warming. The results are displayed on figure 5.8.

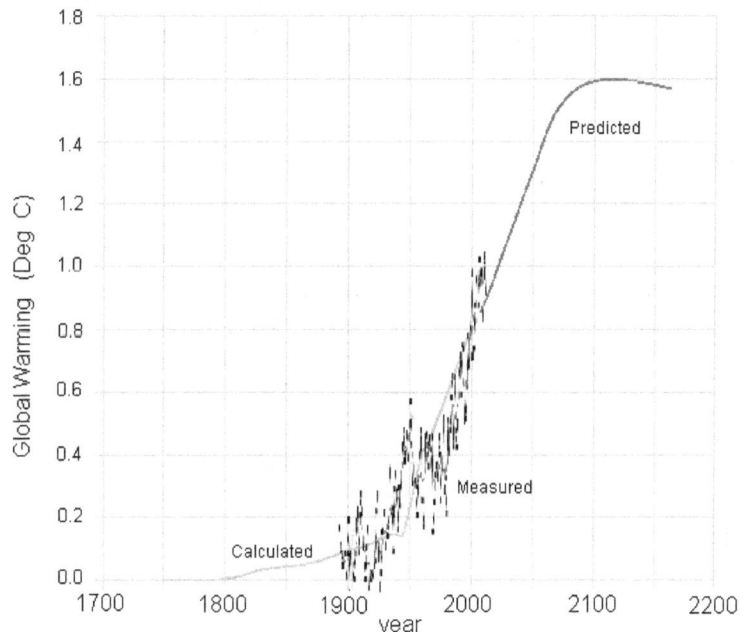

Figure 5.8 Predicted global warming as a result of the atmospheric CO2 concentration of figure 5.7

Calculated from the radiative forcings from the CO2 of figure 5.7, in addition to constant forcings for CH4 and CFC, and a water feedback forcing of 1.0 W/m² per °C warming.

The peak CO_2 atmospheric level of 620 ppm produces a global warming of 1.6 °C above pre–industrial temperatures. This temperature rise is again much smaller than most of the values quoted in the literature. The reason is that the CO_2 has only a weak forcing, which becomes relatively weaker at higher concentrations, and it has been assumed that the other forcing of the trace components remain constant. In particular, most global warming estimates come from complex global models which predict considerably more ' feedback' than the water feedback allowed for here. Feedbacks will become a major preoccupation of chapter 7, as they are so important to get right.

The curve however does suggest that the planet has already warmed half as much as it ever will in the future. The planet has warmed 0.8 °C, the peak will be 1.6 °C – such figures are themselves not unduly alarming.

5.3c) Future sea levels

Sea level changes can be calculated from future global temperature rise, assuming that the majority of sea level change is the result of the density change of water. With an active depth for the sea of 700 m, the expansion can be calculated from the temperature rises given in Figure 5.4 to produce sea level rises shown in Figure 5.9. This suggests the seawater level will rise a total of 180mm due to global warming, of which it has already risen nearly 100mm.

This is not an alarming rise. Admittedly feedbacks and melting ice caps have not entered into the discussion, and these are considered important by the more complex models.

5.3d) Future sea composition

Using $CO_2/HCO_3'/CO_3''$ /H^+/OH' equilibrium data from Zeebe (2001), and seawater *DIC* from the fitted mass transfer

114

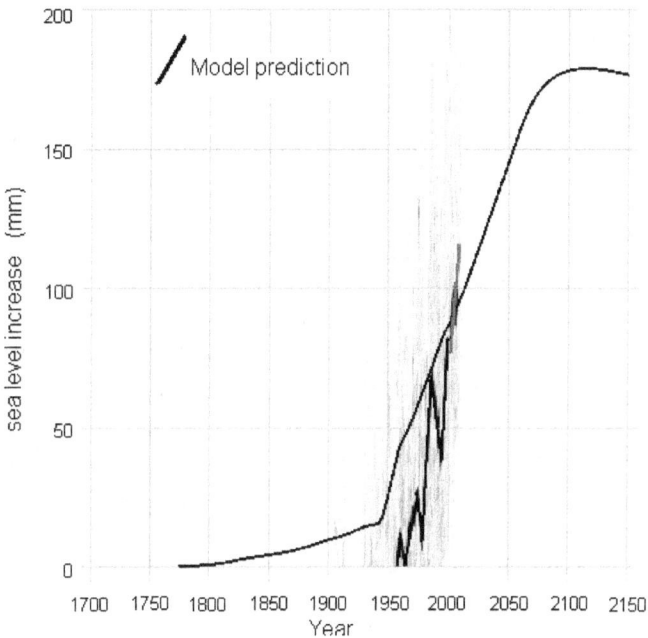

Figure 5.9 Future sea levels as a result of the global warming presented in figure 5.8

This curve simply reflects the change in sea density assuming 700m of the sea attains the global warming temperatures of figure 5.8. As explained under figure 5.5, the model predicts about half the rise that is observed. This suggests the maximum sea water rise might be 150 mm above present levels.

calculations from this study, the future molar concentration of H^+ and hence the pH can be calculated – see the Appendix to chapter 3. This shows a fall from 8.17 to 8.02 over the next 150 years – a fall of 0.15pH units. This model assumes that the anthropogenic CO_2 is distributed evenly over the whole of the ocean depth. Since we are very certain that this not the case, this can be taken as the minimum possible pH change – the change is likely to be greater.

The other extreme is to assume that the surface water will be in equilibrium with the CO_2 in the atmosphere, and this will give an upper limit to future expected pH changes. For an atmospheric CO_2 concentration of 620 ppm, using the Appendix to chapter 3 we can calculate the pH will change from 8.17 to 7.90 – a reduction of 0.27 pH units.

A third possibility is following the suggestion that the alkalinity changes will occur, making the sea behave as though it was equivalent to the fully mixed ocean – giving the change of 0.15 pH units which corresponds to our first calculation..

These pH changes should be compared with the present range of pH measured in seawater, and past levels determined by boron isotope measurements on coral reefs. This is reported in Chapter 4 and measured values of pH are quoted between 7.9 and 8.4.

This CO_2-induced pH fall is not going to recover in the way that the ppm CO_2 in the atmosphere, global temperatures and sea water levels will recover after the fossil fuel burning period of our civilization is over, unless the CO_2 neutralization by the $CaCO_3$ reaction is significant Without this reaction, pH will continue to decrease well after the CO_2 atmospheric levels and global temperatures have peaked.

It is very important to investigate this $CaCO_3$ reaction, as the change in pH of the upper layer of the oceans may be the most damaging result of continued fossil fuel burning.

5.3e) Future Extreme Events

Using the same arguments as section 5.2d, we can calculate the change in probability for extreme events, assuming a Normal Distribution based on the new global temperatures and corresponding new water vapor pressures. As section 5.2d, the 1: 100 probability levels will reduce . Table 5.1 includes the probability levels for 2150, at the peak CO_2 atmospheric concentrations, as well as the 2000 levels which have been observed.

116

Table 5.1   Probability of Extreme Events

Year	1900	2000	2100
Event			
Extreme precipitation	1 in 100	1 in 70	1 in 53
Heavy precipitation	1 in 20	1 in 17	1 in 14
Extreme Heatwave	1 in 100	1 in 65	1 in 40
Heatwave	1.in 10	1 in 8	1 in 6

5.4) Conclusions

5.4a) – on the Predictions of the future

A simple mass transfer model has been found to explain adequately the atmospheric CO_2 levels over the last 100 years from released CO_2 from fossil fuel data. This model also predicts well the observed changes in mean global temperature, sea level and occurrence of extreme events that have been observed over the last century.

Using this model for the prediction of the next 150 years, with a modified 'business as usual' scenario shows:– the CO_2 levels in the atmosphere to peak at about 620ppm after about 100 years; temperatures will rise 0.8 °C above today's temperatures; the pH of the oceans is likely to reduce by between 0.15 and 0.27 pH units by the year 2150, seawater levels will rise 150 mm above their present levels; the probability of extreme events will double compared to 1900, where 1 :100 precipitation and heatwave events in 1900, will become 1: 50 events at the peak of the global warming. The present global warming value of 0.8°C is halfway to the maximum predicted 1.5°C. This means we are already experiencing all the effects in half measure.

This analysis has been carried out with data which is well established, and does not conflict with the IPCC reports. – one exception being that the possibility of various 'feedbacks' predicted by the climatologists general circulation global models (GCM models) have not been included. The GCM models also consider some CO_2 'sinks' to be terrestrial. The simple model used here ignores any long–term storage terrestrial possibilities, assuming the only sink for CO_2 is the sea.

Most surprising are the low levels of atmospheric CO_2 in the future, rising only to 620ppm, even when the fossil fuel usage is trebled. The reason for this is that mass transfer into the sea depends upon a driving force – the CO_2 pressure difference – it takes time to build this up.

5.4b) – on the validity of the model

These predictions have been made using a very simple 'one – dimensional' model This model is fully justifiable on statistical grounds as a means of understanding the mechanisms involved from the measured data available. Because the fit is so good, there is no need to use more complex models to interpret the observed data. This fitted model has been used for predicting the future.

The model is ' trivial' in global modeling terms, whereas weather forecasting models (GCMs) are extremely complex software, and these models can also be used to fit the measured data and predict futures. Normally, these GCM models are the ones used for global warming predictions as they are more complete and are able to predict in much more detail than simple models. Their disadvantage is that they are so complex that it is difficult to understand their results and implications, which is a real disadvantage from a

118

scientific point of view.

The next chapter will describe these more complex models and explain the results found in using them for fitting the measured data and predicting possible futures. The chapter following will discuss the differences in the results from the two types of model.

.

Chapter 5 Appendix

Developing the Mass Transfer Model

Whatever complications there are for the CO_2 mass transfer, it can be expected that the mass transfer will be according to the principle that the rate of transfer is proportional to the driving force difference between phases. This means the CO_2 partial pressure in the atmosphere (p) and the CO_2 partial pressure exerted by the seawater (p*), calculated using Henry's law together with the chemical equilibrium of the seawater system, determine the mass transfer according to the following relationship:–

$$dc/dt = \text{rate of transfer} = Kga\,(p{-}p^*) \qquad \text{tns/year} \qquad (1)$$

where:–

p^* is the equilibrium CO_2 pressure exerted by the seawater ppmv

p is the partial pressure of CO_2 in the atmosphere,   ppmv

Kg is the mass transfer coefficient per unit area   tns /(yr m^2 ppmv)

a is the active surface area of the oceans m^2

Kga is ill defined, being a complex mix of areas;– spray, waves, mixing, wind, and chemical enhancement, far removed from the engineers expectations for Kga of mass transfer theory. It is however a single unknown, which should be constant, and can be used in a CO_2 global transfer model to see how useful it might be. It can be considered a 'lumped parameter' which will also be much affected by the degree of mixing throughout the depth of the sea.

Data is available for the annual production of CO2 from fossil burning from 1700 to 2009 *(CDIAC, 2011)*, $E_r(n)$ (expressed as Giga tons /year as carbon) – see Figure 5.1 (CDIAC).

Data is available for the concentration of CO2 in the atmosphere from 1850 to 2010, *Mauna Loa 2012 for 1850 – 1982, Keeling for 1958 – 2010(Scripps).* – see Figure 5.2 . A CO2 global model based on equation (1) can describe the movement of this CO2 with only one unknown parameter, to see if the model can describe the observed atmospheric CO2 concentration/time trajectory.

Starting in year 1700, with the sea in equilibrium with the long time CO2 level of 280ppmv, each year can be stepped, the amount of CO2 transferred to the sea calculated by equation (1), that not transferred being used to increase the CO2 level in the atmosphere. The CO2 transferred to the sea is added to the total carbon dioxide in the sea, increasing the DIC, and a revised CO2 equilibrium pressure calculated for the next year's transfer. Only one parameter is available for adjustment *(Kga)*, and the aim is to see whether the observed CO2 concentration vs. time profile in the atmosphere can be reproduced by the model. This procedure can be conveniently performed on a spreadsheet.

The equations regulating each annual step are as follows.

mass of air on the planet $M_{air} = 5.35 \times 10^{15}$ tns

total mass of seawater(mean depth 4000m) $M_{sea} = 13.5 \times 10^{17}$ tns

starting CO2 level in the atmosphere $p(1) = 280$ ppmv

starting equilibrium pressure of CO2 in the sea $p^*(1) = 280$ ppmv

starting total CO2 content of the sea $A(1) = 1.29 \times 10^{14}$ tns

Given carbon emission for year n as input data
$E(n)$ tns/ yr C

this translates to $3.667\ E(n)$ tns /yr CO2

Mass Transfer equations:

CO_2 transferred to seawater
$$T(n) = K_g a(p(n-1)-p^*(n-1)) \qquad \text{tns/yr (2)}$$

CO_2 remaining in atmosphere after transfer,

$$A(n) = A(n-1) + E(n) - T(n) \qquad \text{tns (3)}$$

translating this to ppmv,

$$p(n) = A(n)/44/((M_{air})/28 \cdot 10^{-6} \qquad \text{ppmv(4)}$$

new CO_2 content of the sea

$$M_{sea}(n) \;=\; M_{sea}(n-1)+T(n) \qquad (5)$$

new Equilibrium CO_2 Partial pressure

$$p^*(n) = P^*(n-1) + Re. \, P^*(n-1).(T(n)/M_{sea}(n)) \quad (6)$$

Re being Revelle factor, where:

$$Re = \frac{\text{relative increase in } CO_2 \text{ seawater partial pressure}}{\text{relative increase in seawater } DIC}$$

The Revelle Factor was calculated from the CO_2 seawater equilibrium (Zeebe, 2001) prepared on a separate spreadsheet. This factor linked the two spreadsheet calculations. Values around 12 were calculated for the temperatures and compositions involved.

Setting this model as a spreadsheet will predict the growth of p, with time, and this can be compared with the observed data. The magnitude of $K_g a$ can then be adjusted by trial and error to get the best fit to the observed data. The fitted value of $K_g a$ can then be used to predict the future.

Developing the Global Warming Model

The model calculates the CO2 content of the atmosphere at any time, and this must be converted into the anthropogenic temperature rise. We can identify 3 contributions to the rise:

1) the forcing due to the CO2 level in the atmospheres

2) the forcings due to other GHG in the atmosphere

3) a feedback temperature rise from the increase water vapour content of the atmosphere.

As shown in Chapter 2, the forcing above 280ppm CO2 created by the actual CO2 concentration can be represented by the equation:

$$F_{co2} = B(1 - e^{-\beta z \, (p-280)}) \qquad (7)$$

and this can be fitted to the MODTRAN simulations with the parameter values:

$$B = 4.60, \quad and \quad \beta z = 0.0028$$

The forcing due to other GHGs were reported (F$_{other}$) as 1.17 W/m2 in 1995, see table 2.1. Since 2000 these gases seem to have stabilized, before that time they grew steadily from 1960 see figure 4.2 and 4.3. Removing the pre–industrial CH4 forcing leaves 0.91 W/m^2 as the anthropogenic increases in GHG forcing. Hence our model can add 0.91 W/m^2 forcing to the CO2 forcings from 2000, and before that time distribute the forcing proportional to the level between 1960 and 1995.

The total forcings from the GHGs can be summed and the resulting temperature rise calculated from the information in chapter 2, that 3.4 W/m2 gives a global temperature rise of 1 K.

$$\Delta T_f = (F_{co2} + F_{other}) /3.4 \qquad (8)$$

Water vapor feedback f$_{H2O}$ of 0.29 will give a further temperature increase by the formula (explained in chapter 7), with a resulting equilibrium temperature rise Δ Teq calculated by:

$$\Delta \, T_{eq} \; = \; \frac{\Delta \, T_f}{(1 - 0.29)} \tag{9}$$

The equations (7), (8), and (9) can be programmed into the mass balance spreadsheet model to provide a prediction of the global warming to date. The Δ Teq calculated is the model prediction of global warming, which is shown on figure 5.4 together with the actual measured data.

Figure 5.4 shows the model fits the measured global warming well. Once again, the agreement is as close as can be expected, considering the scatter that is present in the measured data.

Table 5.2 is the output from the spreadsheet calculation, and shows the stepwise calculation procedure. The value of K_ga (field A1) is altered by trail an error until there is agreement in the CO_2 partial pressures observed and predicted (column F and G). The value of K_ga of 1.58e+8 tn /(yr ppm) gave the best fit. The table also shows other data of interest that had to be calculated on the way – for instance the annual % CO_2 absorbed by the sea, and the back partial pressure of CO_2 exerted by the seawater.

Table 5.3 is the extension of the spreadsheet calculation into the future, showing how, as the atmospheric CO_2 level increases, the % absorbed by the sea also increases, so producing a stabilizing effect on the attained CO_2 levels.

Table 5.2   Model Fitting   4000m sea depth, no CO_2 neutralisation Model

n	E(m)	T(m)	D	A(r)	p(n)	p(n) obs	% to sea	I	p*(n)	K	DIC	DTeq
A	B	C	D	E	F	G	H	I	J	K	L	M
1.5E+8							1.00	4.0000E+00	280.03	1.2550E+14	1.1800E+01	0.00
1750	3.00E+006	0.00E+00	1.10E+07	2.35E+12	280.6	280.0	0.00	0.0000E+00	280.03	1.2550E+14	2.0373E-06	0.00
1800	8.00E+006	1.03E+07	1.90E+07	2.35E+12	280.1	280.0	35.24	0.0000E+00	280.03	1.2550E+14	2.0373E-06	0.00
1825	4.50E+008	2.61E+07	1.62E+09	2.36E+12	280.2	280.0	1.58	0.0000E+00	280.04	1.2550E+14	2.0374E-06	0.03
1850	4.60E+008	7.87E+08	9.00E+08	2.42E+12	287.7	288.0	46.63	0.0000E+00	280.56	1.2554E+14	2.0380E-06	0.04
1875	5.90E+008	1.13E+09	1.03E+09	2.44E+12	290.8	291.0	52.23	0.0000E+00	281.29	1.2560E+14	2.0389E-06	0.06
1900	1.05E+009	1.50E+09	2.35E+09	2.50E+12	297.8	298.0	38.98	0.0000E+00	282.25	1.2567E+14	2.0401E-06	0.09
1910	1.10E+009	2.45E+09	1.58E+09	2.52E+12	299.7	300.0	60.80	0.0000E+00	282.89	1.2580E+14	2.0421E-06	0.10
1920	1.35E+009	2.65E+09	2.30E+09	2.54E+12	302.4	303.0	53.52	0.0000E+00	283.57	1.2593E+14	2.0443E-06	0.12
1930	1.75E+009	2.97E+09	3.44E+09	2.58E+12	306.5	307.0	46.34	0.0000E+00	284.34	1.2608E+14	2.0467E-06	0.14
1940	1.30E+009	3.50E+09	1.26E+09	2.59E+12	308.0	311.0	73.45	0.0000E+00	285.25	1.2625E+14	2.0495E-06	0.14
1945	1.16E+009	3.59E+09	6.61E+08	2.59E+12	308.4	311.0	84.46	0.0000E+00	285.72	1.2643E+14	2.0525E-06	0.15
1950	1.63E+009	3.58E+09	2.40E+09	2.60E+12	309.8	311.0	59.91	2.7900E+10	286.19	1.2650E+14	2.0536E-06	0.23
1955	2.04E+009	3.73E+09	3.76E+09	2.62E+12	312.0	314.0	49.82	6.4630E+10	286.68	1.2658E+14	2.0548E-06	0.31
1960	2.57E+009	4.01E+09	5.41E+09	2.65E+12	315.3	316.9	42.55	1.1075E+11	286.89	1.2666E+14	2.0561E-06	0.40
1962	2.69E+009	4.48E+09	5.37E+09	2.66E+12	316.5	318.5	45.52	1.1601E+11	287.12	1.2667E+14	2.0563E-06	0.42
1964	3.00E+009	4.65E+09	6.33E+09	2.67E+12	318.0	319.6	42.32	1.2169E+11	287.37	1.2668E+14	2.0564E-06	0.43
1966	3.29E+009	4.85E+09	7.21E+09	2.69E+12	319.8	321.4	40.21	1.2797E+11	287.62	1.2669E+14	2.0566E-06	0.45
1968	3.57E+009	5.08E+09	8.00E+09	2.70E+12	321.7	323.0	38.84	1.3482E+11	287.89	1.2670E+14	2.0567E-06	0.47
1970	4.05E+009	5.34E+09	9.53E+09	2.72E+12	323.9	325.7	35.91	1.4244E+11	288.17	1.2671E+14	2.0569E-06	0.49
1972	4.38E+009	5.65E+09	1.04E+10	2.74E+12	326.4	327.4	35.21	1.5087E+11	288.47	1.2672E+14	2.0571E-06	0.51
1974	4.62E+009	5.99E+09	1.10E+10	2.77E+12	329.0	330.2	35.36	1.5987E+11	288.79	1.2673E+14	2.0573E-06	0.53
1976	4.86E+009	6.36E+09	1.15E+10	2.79E+12	331.7	332.0	35.64	1.6936E+11	289.12	1.2674E+14	2.0575E-06	0.55
1978	5.19E+009	6.73E+09	1.23E+10	2.81E+12	334.7	335.4	35.41	1.7941E+11	289.48	1.2676E+14	2.0577E-06	0.57
1980	5.32E+009	7.14E+09	1.24E+10	2.84E+12	337.6	338.7	36.63	1.8991E+11	289.86	1.2677E+14	2.0580E-06	0.59
1982	5.11E+009	7.54E+09	1.12E+10	2.86E+12	340.3	341.4	40.24	2.0034E+11	290.26	1.2679E+14	2.0582E-06	0.61
1984	5.28E+009	7.90E+09	1.15E+10	2.88E+12	343.0	344.6	40.79	2.1074E+11	290.67	1.2680E+14	2.0585E-06	0.63
1986	5.61E+009	8.27E+09	1.23E+10	2.91E+12	345.9	347.4	40.20	2.2163E+11	291.11	1.2682E+14	2.0587E-06	0.65
1988	5.97E+009	8.66E+09	1.32E+10	2.93E+12	349.1	351.6	39.57	2.3321E+11	291.58	1.2683E+14	2.0590E-06	0.67
1990	6.15E+009	9.08E+09	1.35E+10	2.96E+12	352.3	354.4	40.28	2.4532E+11	292.06	1.2685E+14	2.0593E-06	0.70
1992	6.18E+009	9.51E+09	1.31E+10	2.99E+12	355.4	356.4	42.01	2.5765E+11	292.57	1.2687E+14	2.0596E-06	0.72
1994	6.29E+009	9.93E+09	1.32E+10	3.01E+12	358.5	358.8	43.01	2.7012E+11	293.10	1.2689E+14	2.0599E-06	0.74
1996	6.55E+009	1.03E+10	1.37E+10	3.04E+12	361.8	362.6	43.04	2.8297E+11	293.65	1.2691E+14	2.0603E-06	0.76
1998	6.64E+009	1.08E+10	1.36E+10	3.07E+12	365.0	366.6	44.22	2.9615E+11	294.23	1.2693E+14	2.0606E-06	0.78
2000	6.75E+009	1.12E+10	1.36E+10	3.10E+12	368.2	369.5	45.18	3.0954E+11	294.83	1.2696E+14	2.0610E-06	0.79
2002	6.98E+009	1.16E+10	1.40E+10	3.12E+12	371.6	373.2	45.31	3.2327E+11	295.46	1.2698E+14	2.0614E-06	0.81
2004	7.78E+009	1.20E+10	1.65E+10	3.16E+12	375.5	377.5	42.14	3.3804E+11	296.11	1.2700E+14	2.0617E-06	0.82
2006	8.35E+009	1.25E+10	1.81E+10	3.19E+12	379.8	381.9	40.97	3.5417E+11	296.79	1.2703E+14	2.0622E-06	0.84
2008	8.75E+009	1.31E+10	1.90E+10	3.23E+12	384.3	385.6	40.88	3.7127E+11	297.50	1.2705E+14	2.0626E-06	0.86
2010	9.01E+009	1.37E+10	1.93E+10	3.27E+12	388.9	385.6	41.51	3.8903E+11	298.25	1.2708E+14	2.0630E-06	0.88
2012	9.28E+009	1.43E+10	1.97E+10	3.31E+12	393.6	389.8	42.09	4.0732E+11	299.03	1.2711E+14	2.0635E-06	0.90

Table 5.3 Model Prediction   4000m sea depth, no CO2 neutralisation Model

A	B	C	D	E	F	G	H	I	J	K	L	M
2010	9.01E+09	1.37E+10	1.93E+10	3.27E+12	388.9	389.8	41.51	3.8903E+11	298.25	1.2708E+14	2.0630E-06	0.88
2012	9.28E+09	1.43E+10	1.97E+10	3.31E+12	393.6		42.09	4.0732E+11	299.03	1.2711E+14	2.0635E-06	0.90
2014	9.56E+09	1.49E+10	2.01E+10	3.35E+12	398.4		42.62	4.2475E+11	299.85	1.2715E+14	2.0642E-06	0.91
2024	1.10E+10	1.56E+10	2.47E+10	3.60E+12	427.8		38.62	5.2752E+11	304.12	1.2731E+14	2.0667E-06	1.02
2034	1.26E+10	1.95E+10	2.68E+10	3.86E+12	459.7		42.16	6.4571E+11	309.56	1.2750E+14	2.0699E-06	1.13
2044	1.45E+10	2.37E+10	2.96E+10	4.16E+12	494.9		44.50	7.8163E+11	316.27	1.2774E+14	2.0737E-06	1.24
2054	1.67E+10	2.82E+10	3.31E+10	4.49E+12	534.3		46.03	9.3793E+11	324.44	1.2802E+14	2.0783E-06	1.35
2064	1.92E+10	3.32E+10	3.74E+10	4.87E+12	578.7		47.02	1.1177E+12	334.28	1.2836E+14	2.0837E-06	1.46
2074	1.73E+10	3.86E+10	2.48E+10	5.11E+12	608.3		60.86	1.3004E+12	346.09	1.2874E+14	2.0900E-06	1.52
2084	1.56E+10	4.14E+10	1.57E+10	5.27E+12	626.9		72.53	1.4648E+12	359.20	1.2916E+14	2.0967E-06	1.56
2094	1.40E+10	4.23E+10	9.10E+09	5.36E+12	637.8		82.29	1.6127E+12	373.10	1.2958E+14	2.1036E-06	1.58
2104	1.26E+10	4.18E+10	4.45E+09	5.41E+12	643.0		90.39	1.7459E+12	387.37	1.3000E+14	2.1103E-06	1.59
2114	1.14E+10	4.04E+10	1.24E+09	5.42E+12	644.5		97.02	1.8658E+12	401.69	1.3040E+14	2.1169E-06	1.60
2124	1.02E+10	3.84E+10	-8.95E+08	5.41E+12	643.5		102.39	1.9736E+12	415.78	1.3078E+14	2.1231E-06	1.59
2134	9.20E+09	3.60E+10	-2.25E+09	5.39E+12	640.8		106.66	2.0707E+12	429.47	1.3114E+14	2.1290E-06	1.59
2144	8.28E+09	3.34E+10	-3.04E+09	5.36E+12	637.2		110.00	2.1581E+12	442.58	1.3148E+14	2.1344E-06	1.58
2154	7.45E+10	3.07E+10	-3.43E+09	5.32E+12	633.1		112.55	2.2367E+12	455.03	1.3179E+14	2.1394E-06	1.57
2164	6.70E+009	2.81E+10	-3.55E+09	5.29E+12	628.9		114.43	2.3075E+12	466.74	1.3207E+14	2.1439E-06	1.57
n	E(n)	T(n)		A(n)	p(n)	p(n) obs	% to sea		p*(n)		DIC	DTeq

The model coding, in .XLS format, can be obtained from the author chemeng@btinternet.com, or his website

References

Archer, D., Kheshgi, H., Maier–Reimer, E., *Multiple timescales for neutralization of fossil fuel CO₂*, Geophysical res, letters, 24, 4, 405–408, Feb 15th 1997

CDIAC, Carbon Dioxide Information Analysis Center, http://cdiac.ornl.gov

CDIAC, Boden, T., Marland, G.,Andres, B.,Trends on line, *Global CO₂ Emissions from Fossil–Fuel Burning, Cement Manufacture, and Gas Flaring: 1751–2008*, 2011.

IPCC(2007) *Fourth Assessment Report: Climate Change (AR4), The Physical Science basis*

Lindzen, R. S.,and Choi, Y., (2009), *On the determination of climate feedbacks from ERBE data*, Geophysical Res. Letters, 36, L16705 1–6

Lu Q, (2010), *What is the major Culprit for Global Warming CFCs or CO2?*, J of Cosmology, 8, 1845–1862

Mauna Loa, Pieter Lans, *datafile, ESRL*, 2012

Manabe, S and Wetherald, T., (1975) The effects of doubling CO2 concentrations on the climate of a general circulation model, J of the Atmospheric Sciences, 32, 3 – 15

Min, S., Zhang, X., Zwiers, F.W., and Hegert, G. C., (2011), Human contribution to more intense precipitation extremes, Nature, 470, 378–381

Myhre, G., Highwood, J. H. Shine, K.,and Stordal, F., (1998), *New estimates of radiative forcing due to well mixed greenhouse gases*, Geophysical Res. Letters, 25, 2715–2718

Ramanathan, V. (1998) Trace gas Greenhouse effect and Global warming Underlying principles and Outstanding issues. Ambio,

27,187–197.

Sabine, C.L., Feely, R. A., Gruber, N., Key, Lee, K., R. M., Billister, J. L., Wanninkhof, R., Wong, C. S., Wallace, D. W. R., Tilbrook, B., Millero, F. J., Peng, T., Kozyr, A., Ono, T., Rios, A. F., *The Oceanic Sink for Anthropogenic CO2,* Science Vol. 305, no. 5682 pp. 367–371, 16 July 2004

Scripps Institute of Oceanography, Mauna Loa Observatory, Hawaii, http://scripps.ucsd.edu/

Zeebe R.E., and Wolf–Gladrow D., (2001) CO2 *in seawater:equilibrium, kinetics and isotopes*, Elsevier, Amsterdam

Chapter 6 Global Climate Models

Climate models of the planet have been well developed for weather forecasting, and these forecasting models have been extended to incorporate the changes in radiation caused by Green House Gases (GHGs) to see the effect on future weather patterns. Many of the predicted consequences of global warming made by these Global Circulation Models (GCMs) are alarming, so it is important to understand why these models predict as they do.

Weather patterns used to be considered to be fairly stable ('tomorrow's weather will most probably be the same as today') modified by random events which produce step changes in conditions. There is the illustration from catastrophe of theory ' when a butterfly flaps its wings in Patagonia, this leads to a hurricane in the Caribbean ' which suggests that the weather can be considered as being influenced by minor, quite random events which determine final outcomes. But weather forecasting is now much more of a science than these illustrations suggest.

It is possible to explain why air movement occurs and possible to determine how air, once moving, will move in directions determined by the forces involved . Air and sea movements might be turbulent, but does not mean that they cannot be expressed mathematically.

6.1) Basic Concepts

Consider a spinning globe. Now radiate this globe with distant radiant heat source, not quite perpendicular to its spinning axis.

Now surround this globe with a thin layer of gas — our atmosphere. This situation was studied by Bjerknes who, in 1904, was the first to study such a system mathematically. Physically, we can see that the radiation will heat the globe most at its equator and will not warm the poles at all because the incident angle for the radiation becomes so shallow near the poles. The poles receive no radiation and so remain cold. The atmosphere at the equator is heated by the hot surface but remains cold at the poles. Since it is a gas, its density will change and the gas will rise at the equator, and fall at the poles. This will introduce circulation patterns, as shown by figure 6.1

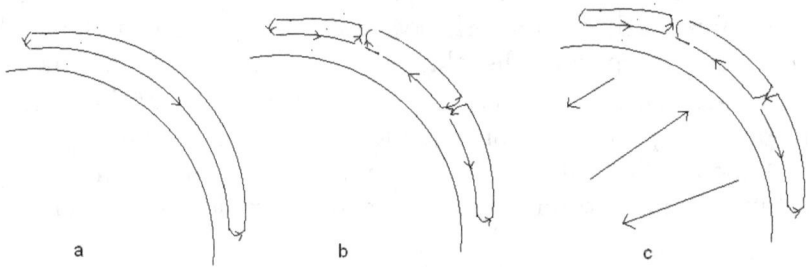

a   b   c

Figure 6.1a Single circulating pattern; 6.1b 3 loops; 6.1c Coriolis effect
The basis of all GCMs is the motion of unevenly heated fluids on
the surface of the globe. The air is hot at the equator, and cold at
the poles. The hot air rises at the equator and falls at the poles,
The single loop 6.1a) is not stable, and the triple loop 6.1b) is the
basis for our winds system. As stationary air moves south from the
poles around the spinning globe it appears to be an east wind –
6.1c). the second loop produces a west wind and the third loop by
the equator produces a weaker East wind . Hence the planet's
Trade Winds are born. All can be calculated, and this is the starting
point for GMCs.

One possibility is for the gas to have one circulation loop, shown by figure 6.1a, but a second possibility is to have 3 loops – see figure fig. 6.1.b . There is no possibility of having 2 loops as this requires the cold gas at the poles to be rising, which will not happen. On our Earth, the stable circulation pattern has 3 loops. Now consider what happens when cold air at the pole circulates towards the equator. The angular velocity is zero at the pole, so as it moves it will be moving backwards relative to the spinning globe surface which has a velocity away from the poles. Looked at from a point on the globe surface, this circulating air is moving backwards – it is a wind. This is known as the Coriolis Effect suggested by G. Coriolis (1792 – 1843).

The direction of the wind is determined by the direction of the spin and the direction of the circulation pattern of the atmosphere. From figure 6.1b with its 3 circulation loops, there will be 3 winds in each hemisphere. This gives two East winds and a West wind in the northern hemisphere, as shown on figure 6.1c) These are the trade winds, which were known in the 17th century and were used by sailing ships, knowing that, by catching the right wind, one could travel most easily in the required direction.

Now we have 3 winds in each hemisphere; in the northern hemisphere a West wind with an East wind on each side. The result is to give spinning motions to the streams, because at the interface of the opposing streams, the air is being dragged in opposite directions which induces a spin. We therefore have a picture of the wind crossing the northern hemisphere with an anticyclone spin, crossing in a predominantly north—easterly direction. This westerly wind over the sea will drag the surface of the sea, results in pulling the surface water westerly. This becomes the 'gulf stream' in the northern hemisphere, and correspondingly becomes the El Nino and Al Nina in the southern hemisphere – between South America and Australia. The ocean currents are dragged by movements along the surface, and the corresponding return cycle is deeper in the ocean. Hence, at the beginning of the stream, cold water emerges from the depths and a heated surface stream travels into colder regions . These currents greatly affect weather patterns.

Our description now includes gas streams traveling along the surface of the ocean, rising into the higher levels of the atmosphere as it warms. This moving air collects water by evaporation as it moves across the sea. As it rises to higher levels of the atmosphere the temperatures falls and the air saturates with water, leading to cloud formation. If further cooling occurs, the quantity of water that can be held in the cooled air is considerable less than it held when the air was warm. The differences will result in precipitation. The magnitude of the difference gives the intensity of the precipitation.

As air rises the temperature falls because the pressure reduces, and to maintain energy balance the temperature must lower. This is called the adiabatic lapse rate — adiabatic because there is no total energy change of the gas. The temperature of the air can be calculated as it rises using this lapse rate and this determines both its ability to hold water and to define whether the water precipitated will be rain or snow.

This whole system is driven by the uneven solar incoming radiation — uneven because the sun is directly overhead at the equator and not visible at the poles; uneven because the spinning axis is not exactly perpendicular. Hence this geographically uneven radiation, plus the day/night change due to the spinning, causes temperature imbalances to occur and the weather patterns to develop. The important part of any GCM is the radiation modeling which determines what part of the globe surface receives energy, when and how much. The outgoing radiation is also an important aspect as this maintains the global heat balance.

Equally as important as the description of the momentum and velocities of the atmosphere, is the transfer of heat which is contained in these flows. Heat transfer from the heated surface to the air creates the temperature changes which generate the movement; energy transfer into the latent heat associated with the vaporization of water transfers heat from sea to atmosphere, and the liberation of latent heat into the atmosphere as water vapor condenses into rain provides enormous energy drivers for the climate.

The surface sea temperatures (SSTs) can be calculated from consideration of radiation, evaporation, and ocean currents. When sea temperature approaches $0\,^\circ$ C it can be assumed that sea ice will form. Similarly, when temperatures move above zero, it means that sea ice will be melting. In this way the models can follow the annual expansion and contraction of arctic sea ice.

The rainfall, calculated from the water balance in the atmosphere, falls on land, saturates the land and finally drains into rivers before returning to the sea, so completing the water cycle. The models know the saturation state of the soil and so can determine the likelihood of flooding when rain on saturated ground results in water runoff instead of absorption. This enables the models to forecast the likelihood of flooding.

The GCM models can handle most of the features required for weather forecasting by making these physical description of our climate into a mathematical model. The globe surface is divided up into a large number of segments – see figure 6.2. The atmosphere is divided into a number of layers, for instance 20 layers between planet surface and Top Of Atmosphere (TOA). So the whole atmosphere is divided into packets at specific geological locations and height. The UK Hadley Centre model for instance has a grid of 96 x 72 points for the surface of the Earth, with a 19 level atmosphere. Since each segment is described by four variables, — pressure, velocity, temperature, humidity, — there are more than 500,000 variables in the model.

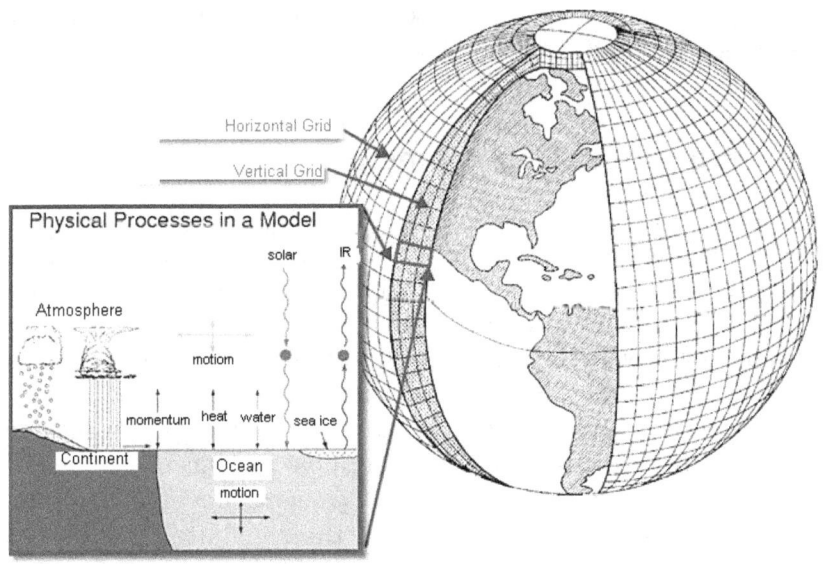

Figure 6.2 Dividing the Globe into Segments
GCMs divide the atmosphere into a number of segments eg 100 x 70 x 20 high = 140,000 segments, and each segment goes through a calculation of the physical processes to progress the heat, momentum, and humidity values by one time step.

The conditions in the segments adjoining each segment affect the conditions of each segment, which changes the values of the variables a little, at each time step, using the Navier—Stokes fluid flow equations. These changes are integration over time, for every segment The result is the slowly developing weather patterns so familiar to weather forecasting animations.

The procedure is best done using a form of coordinate geometry because all the calculations are concerned with the surface of a sphere. The grid—based integration is a mathematics procedure called *finite differences*, used together with *integration* and *non—linear equation solving*. On top of this mathematical structure is added a radiation model such as MODTRAN4, described in chapter 2, which handles both incoming and outgoing radiation. This enables the heat balance and hence surface temperature, as well as temperatures in the various layers of the atmosphere to be determined. In this way the absorption and emission of IR by the GHGs is incorporated into the model.

6.2) Global Model Development

These models were originally developed for forecasting purposes, to be able to predict from a present situation how the weather patterns will change in the next few days. As these methods improved they have been able to give reliable forecasts for a week or two. These models have been modified for global warming predictions, still to consider the whole globe, but in rather less detail, with the emphasis on radiative forcings produced by increasing GHGs. Models of different complexity were developed, from the relatively simple 'Earth Models of Intermediate Complexity' — EMICs — to a series of Global Circulation Models – GCMs. Firstly the Atmospheric Global Circulation Models —AGCM—, then the Oceanic Global Circulation Models, the — OGCMs —, and incorporating this we arrive at the present state of the art with the AOGCMs, which include circulation models of both the atmosphere and the oceans. These models become evermore complex, and it becomes more difficult to interpret the meaning of the results, and the physical arguments behind them. Taking weeks of computing on the world's most powerful computers, they are nearly as complex as the climates they are simulating.

All the models have similar backgrounds because older models have been used for the development of new generations. As one particular difficulty is overcome by clever coding, this may be transferred to other models to save on coding development costs. The result is, these models cannot be considered as independent solutions to the prediction problem, as if you go far enough back there will be many items in common. They are all just one possible solution for the future climate, with some detailed differences and different selections of parameter values.

On top of this basic structure it is possible to attach a wide variety of detailed models, to incorporate as much science as is known – cloud models, mass transfer models, topography models to correct for weather patterns induced by hills and mountains, etc, etc. This detailed model is a receptor for all scientific information, but it needs the correct parameters to be found to make the extra refinements usable. The resulting calculation load is enormous, requiring the biggest computers available, even though clever means for reducing the computational load are incorporated to get simulation times within reasonable limits.

After all, the model takes one tiny segment of the earth's surface, and calculates a radiation model for that segment over a multi–layer atmosphere using a line–by–line or equivalent spectroanalysis procedure to investigate the absorption of each wave length in each layer of each segment. This is repeated until an overall temperature balance is obtained for that column of air. Each parcel of the atmosphere is then analyzed numerically to determine the changes that will occur within it as a result of its momentum and the influences of the adjacent packages. This occurs for each of the packages and may involve iteration to get a solution. The procedure then repeats for the next time step. Hence to predict the weather patterns for 5 days is a long calculation; prediction for 100 years is what is needed.

6.3) Tuning and Parameter Fitting

A mathematical model relates independent variables (input data) to dependent variables (results) by equations and parameters. In our case the independent variables are those that define the present weather, and the dependent variables define the weather after the next time step, which is the object of the calculation. The parameters are constants and fixed numbers that are required by the equations.

Sometimes, these parameters are know because the represent known physical values – such as   Stefan's constant, but most parameters have   be estimated from measurement. Because the model results are so sensitive to the values of these parameters it is necessary to adjust their value even when their value is fairly well known. This adjustment to measured information is sometimes call parameter fitting, but in the meteorological circles it is often referred as model tuning. The more complete the model, the more variables are involved, but also there are very many more parameters. In varying the values of the parameters to fit the measurements, a statistical process called regression, it is usually found the more complex the model the more difficult it is to determine the values of these parameters with certainty. The worst problem is that of 'correlation' which means that one set of parameters can give an equally good fit by the model to the measured data as a different set. In fact there can be a wide range of parameter values which give equally good fits as long as other parameters values are altered correspondingly. The better the model formulation, and the wider the range of measurements, the more likely it is that the parameters can be given precise values.

The problem with correlation is that, although the model can reproduce the measured data, whenever the model is used outside the range of measurements, then the results will depend upon which set of parameters has been used.

This means the model is not useful for predicting new conditions, it is just a convenient way of summarizing lots of experimental data.

The best method of removing this problem of correlation and

determining which set of parameters is the correct one is to have a wider range of measurements, covering variations in more variables. In the laboratory this is easily possible using a technique know as 'experimental planning'. In meteorology, one can only work with the recorded weather data. There is no opportunity to plan experiments.

Since the first mathematical descriptions of the climate were made in 1922 by Richardson, the GCM model have been very well tuned and developed so that their predictions of future weather from the past weather is very impressive, as shown by their 5 day forecasts. Longer term forecasts are less accurate but continued development of the models will no doubt extend the accuracy of the longer range forecasts.

Using these models for prediction of global warming requires the parameters to be fitted from historic, global warming data. This is not as simple as it seems, because the measured data has been affected by a multitude of minor influences – volcanoes, variations in incoming radiation, El Nino – and the results from the GCMs themselves are variable. Figure 6.3 shows detailed fits of two models to the measured data.

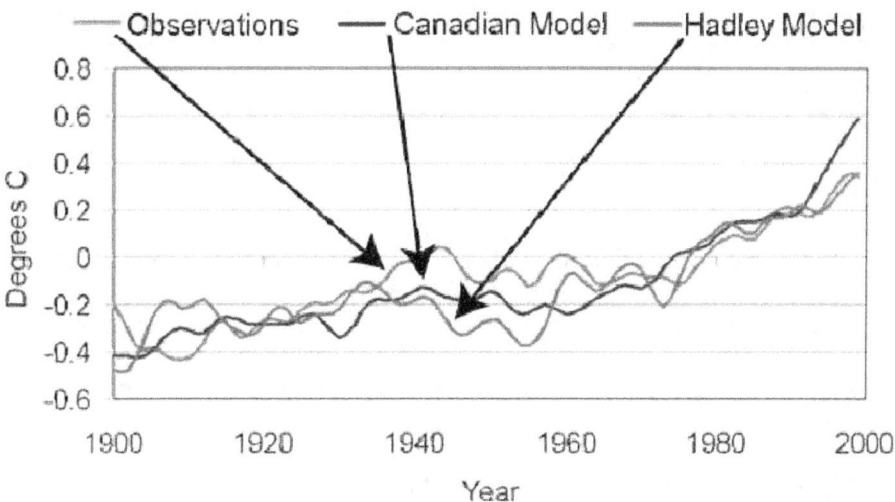

Figure 6.3 Different Models fitted to observed global warming data
It is no easy matter to decide whether the varying observed curve is well fitted or not by the varying results of the complex GCMs. Two examples are given here.

All GCM models must be compared with the available global warming data, for them to be credible, and also to provide them with a defined starting point. IPCC (2007) have compared a range of mean global warming predictions from different models, and these are shown together on figure 6.4. It can be seen that they all fit the measured data region (1970 – 2010) as well as each other, and as good as the fit in figure 6.3 and equally well as the simple analysis shown in chapter 5.

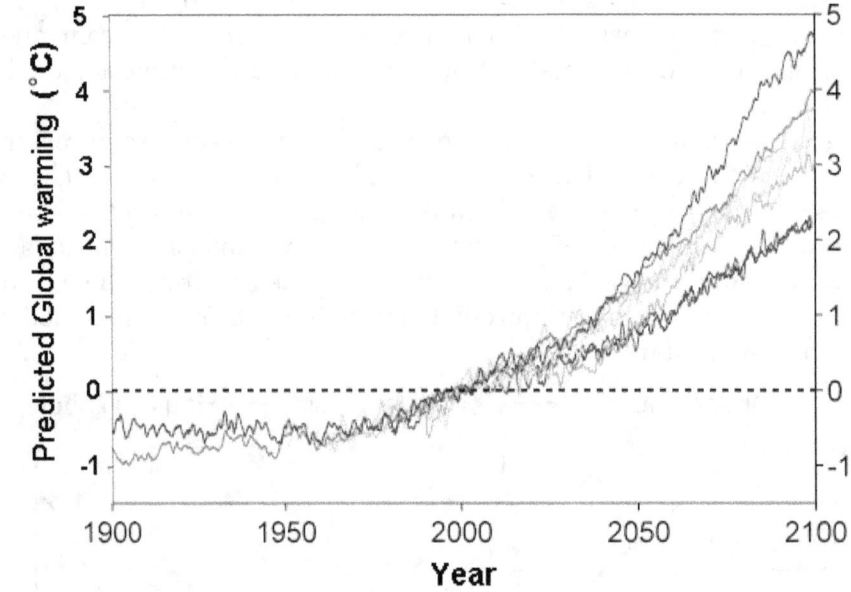

Figure 6.4 Comparing Predicted Mean Global Warming from eight different GCMs.

All 8 models agree and fit the measured temperature data well from 1970 – 2010.— but, because different correlations between the large number of parameters, the models diverge beyond the fitted region.

This is not surprising, because the more complex the model the more parameters it has for tuning and the better the fit to any measured data. The 'noise' associated with the global warming curve means that it is not possible to make comparisons of the fits of each model to determine which has the edge on accuracy.

6.4) Predicting Future Global warming with the GCMs

Predictions of the amount of global warming to occur over the coming 20 years are similar for all models, but beyond that time, divergence occurs between them. This is clearly seen in figure 6.4. This divergence is brought about by the problem of correlation, where all sets of parameters predict near futures reasonably accurately, but when moving further beyond the fitted region, it depends on what set of equally good fitted parameters is chosen for the long term trends. Predictions for fifty years hence vary wildly; from a warming of 1.5 °C which is tolerable to 5 °C which will cause major problems for us all. Hence the greatest need in the global warming study is to resolve the difference between the different models and their different sets of parameters, to determine which is more likely to be the correct future prediction. One approach by Knutti et al (2008) is to consider all the different model forecasts and try to apply Bayesian statistics to be able to report a range of results with attached probabilities. This produces a wealth of data, but one wonders what can be done with it. We have only one planet, we need to know what the future will be. Bayesian statistics, attaching subjective probabilities, is fine when there are 10 planets and 10 different futures, to help get the best mean result. But when there is only one future, decision criteria should be based on minimizing catastrophe, not maximizing subjective probability. There is no easy way other than simply getting the right answer!

Since the models work with a very detailed mathematical description of the surface of the globe, the different layers of the atmosphere and the depths of the oceans, these models do calculate vast amounts of detail, though the result most often reported is the mean global temperature rise.

The main strength of these global models is that they look at the whole planet regionally, and also they consider heat transfer in detail. This means they show regional results in ways that the simple models cannot possibly do. Some of the GCMs most interesting results suggest that the major temperature rises will be in the Arctic and in the north Northern hemisphere, see figure 6.5. Land masses will see greater rise than the sea, and much of the oceans will experience very little temperature rise. Sea ice in the Arctic will reduce considerably, and this will show itself in the increase in temperature to be expected in the Arctic.

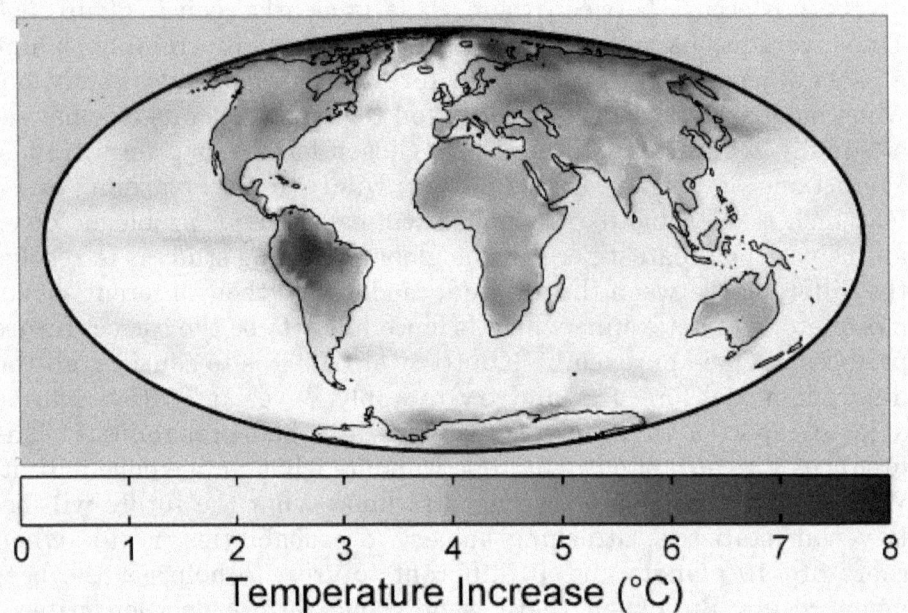

Temperature Increase (°C)

Fig 6.5 Predicted Regional Global Warming

The regional nature of the GCM models gives them the ability to indicate which parts of the planet will be most affected by global warming. The Arctic regions will have the greatest temperature rise. Land masses will experience a greater temperature rise than the oceans. In particular, the models predict that the present arid areas will get drier, because although there will be generally more rainfall, the dry areas will experience more evaporation from the land surface.

Predictions of rainfall are generally in line with the increased vapor pressure at the higher global temperatures, but less rainfall is expected on the great land masses – where there are already deserts. A particular disadvantage of global warming is that it will increase evaporation of water from the soil, and this can intensify arid conditions even when there is more rainfall overall. This sort of detail is brought out well by the use of GCMs.

6.5) Conclusions

The GCM models are impressive in their ability to contain so much Climate Science and with it to be able to predict short term weather patterns with considerable accuracy. When they are developed to investigate the effect of CO_2 and global warming, they are using Climate Science as much as possible to indicate what the implications could be. But since these models cannot be properly tuned because the available observation data is inadequate to remove correlation, it seems not possible to get definitive *quantitative* results for the prediction of future global warming. Different models give different possible global warming values – between 1.5 and 5 °C. Along with this wide variation comes uncertainties in other future conditions to be expected with the warming. This range of results is too wide for sensible decision-making. It is unlikely that more runs with more sophisticated GCMs will overcome this correlation problem – the results may converge with more work, but this may be because the codings become more and more similar (Paltridge, 2009).There is a need for some independent verification of the GCM results (Lindzen, 2009).

Ways out of this impasse could be to ;—

a) approach the problem from a completely different direction;

b) analyze all the parameters involved in the correlation to identify them more accurately.

c) to get more measured data;

The simple model described in Chapter 5 is an attempt at approaching the problem in a different direction. In the next chapter we will discuss the differences between the GCM and the simple model results to try to explain the wide range in the predicted futures. This next chapter will also discuss the parameters causing the main correlation problem to identify them more accurately.

To look at paleontology may help us find more data. This is the subject of chapter 8.

References

IPCC, Working Group I, to the Fourth Assessment Report of the Intergovernmental Panel on Climate Change,(2007), *The Physical Science Basis,* Solomon, S., D. Qin, M. Manning, Z. Chen, M. Marquis, K.B. Averyt, M. Tignor and H.L. Miller (eds.)Cambridge University Press,

Lindzen, R. S.,and Choi, Y., (2009), *On the determination of climate feedbacks from ERBE data,* Geophysical Res. Letters, 36, L16705 1—6

G W Paltridge, (2009), *The Climate Caper,* Connon Court publishing,

Further Reading

Knutti, R., Allen, M. R., Friedlingstein, P., Gregory, J. M., Hegerl, G. C., Meehl, G. A., Meinshausen, M., Murphy, J. M., Plattner, G. K., Raper, S. C. B., Stocker, T. F., Stott, P. A., Teng, H., Wigley, T. M. L.(2008), *A review of Uncertainties in Global Temperature Projections over the Twenty-first Century,* Journal of Climate Science, 21, 2651.

Linacre, E., Geerts, B., (1997) *Climates and Weather explained,* Routledge, London,

Washington, W. M. and Parkinson, C. L., (2005), *Introduction to 3—Dimensional Modeling,* University Science Books, US

Chapter 7   Comparing the Models

7.1) Model Structure

So the last two chapters show that the predictions from the simple averaging model and the complex GMC models are different. The Simple Model produces an expected global warming of a further 0.8 °C, which is tolerable, whereas the complex GMC models predict the future temperature rises somewhere between 2 and 5°C, which would be intolerable. In deciding what remedial actions society should take, and how quickly they should act, it is essential to know which of these predictions is more likely to be correct.

So we need to compare the two types of model, to understand how these differences have arisen. Let us start by listing the differences in the two types of model and then try to discover why there is the wider range and higher global warming predictions given by the GCMs. This list is given in Table 7.1.
Out of this, we can try to find what are most likely causes for the wider spread and greater severity of results of the GCM models. Remember, we are looking for multiple factors that will exhibit correlation ie they compensate for each other for the fitted range years 1950–2000, but diverge strongly beyond 2030.

There are 4 areas where the differences are most likely:–

    a) Arctic sea ice
    b) Water vapor effects
    c) Sea temperature lags
    d) Absorption of CO_2 by the sea

Let us look at these individually:

Table 7.1
List of Model Differences

Simple 1–dimensional Model	4–dimensional GCM Models
Model structure	
Single point mean model	Multi–segment geographic model
Equilibrium model	Dynamic model, integrating with time
Equilibrium heat balance	Heat transfer rate model
Ignores ice formation	Considers ice cap changes
Ignores cloud formation	Models clouds
Ocean considerations	
Assumes deep, mixed ocean	Detailed ocean circulation model
Ocean at global mean temperature	Slow temperature rise in whole ocean (lag)
No Land CO_2 sink considered	Some CO_2 absorbed by the land
Engineering mass transfer and chemical kinetic modeling, based on lumped parameters	Mass transfer and kinetic modeling, may be different from simple model
Emission Forecasts	
CO_2 future concentrations predicted by scenario developed in Chapter 5	Various CO_2 futures considered; IPCC scenario 'A1B' is closest to scenario developed in Chapter 5

7.1a) Arctic Sea Ice

As the global temperature increases the temperature of the sea changes and the sea ice melts, The albedo (reflectivity) of the ice is greater than the albedo of the water, and so more sunlight is absorbed, so giving more global warming. The simple models completely ignore the difference in temperature across the globe and so have no analysis for sea ice. Sea ice is important in the planet heat balance because the reflection of sunlight from ice is of the order of 80% of the incoming isolation. Water with a low sun has a reflection of about 38% (Linacre, 1997) This is called the 'albedo' of the surface. Hence. if ice has formed, it reflects more light and so will induce more cooling. As ice melts, the sea replacing the ice will cause further heat to be absorbed, and so further temperature rise to occur. The GCM models predict areas of ice coverage, and report considerable temperature changes in the arctic regions, no doubt due to this effect.

But now look at the quantities involved. The sea ice has shrunk by 30% in the last 30 years. Figure 7.1 shows the 2 average ice cap sizes, for 1979 and 2007. The irregular patterns of the sea ice on this figure can be averaged giving a mean ice boundary in 1979 at 78° latitude, and in 2007 at 82° latitude. These ice limits can be redrawn as figure 7.2, which shows the amount of incidental radiation involved by the fraction of ice coverage. It is quite small.

When one looks at the results of the GCM global temperature distribution results, it is clear that the Arctic temperatures have risen the most. This suggests that the Arctic ice changes have allowed the Northern temperatures to increase considerably, so the effect on the rest of the globe will be correspondingly less.

It is difficult to see how the Arctic sea ice can be contributing significantly to global warming. It does appear that though the ice cap arguments are well founded, the magnitude of the effect is small.

Melting ice caps are likely to be doing more good than bad, opening up some northern regions of the planet to cope with the ever expanding human population.

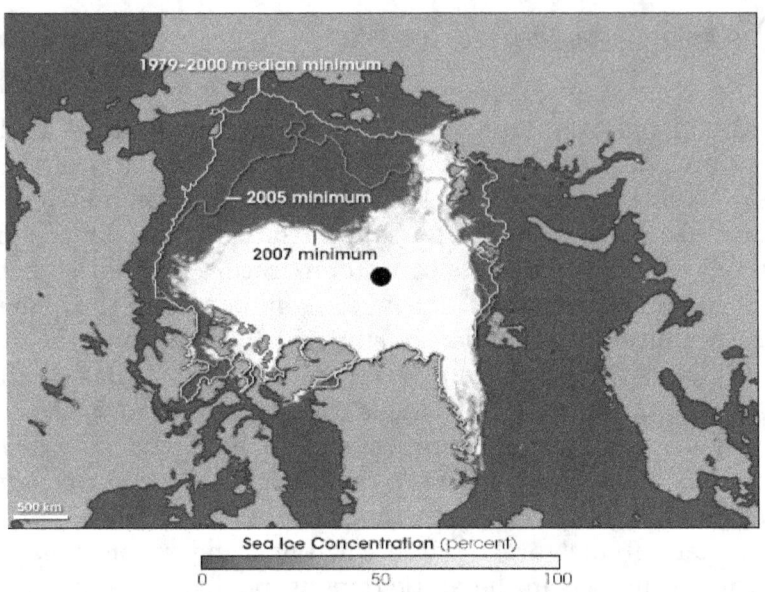

Figure 7.1 Arctic Ice Decline

Over 30 years the sea Ice coverage has declined by 30%. From this image it is possible to calculate the change in the increased incoming radiation caused by the albedo change from ice to water

Figure 7.2 The Arctic ice shadow

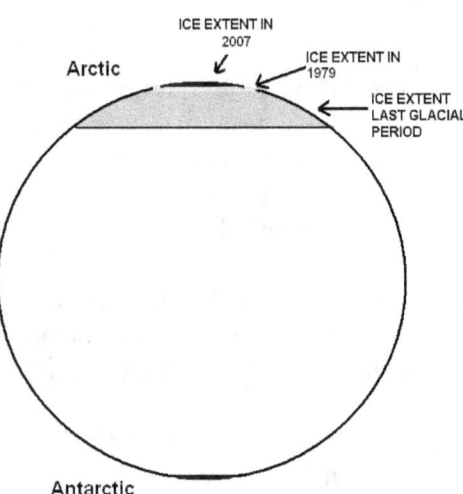

With the maximum ice extent reaching 78 degrees latitude in 1979, and 82 degrees in 2007, the amount of direct sunlight involved on the ice surfaces is negligible. At the height of the last Ice age the ice reached 48 degrees, which will have a marked effect on the absorbed radiation from the Sun. The Antarctic area changes very little; being open ocean and sea currents prevent the formation of sea ice

7.1b) Water Vapor Effects

As the global temperature increases, water vapor pressures increase, and we can expect a 6% rise in all events associated with water vapor per 1 °C rise in global temperature. This is taken into account in both the Simple and the GCM models, but the GCM models do show the contribution of water to temperature rise is very much higher than the Simple Model, which uses the reasoning discussed in Chapter 2. Some GMC results suggest that the water will contribute as much as the CO_2 – ie doubling the effect of the CO_2, whereas Chapter 2 suggests the water contributes only 29%. This increased importance of water vapor in the GCM models may be the result of the more thorough analysis of the complex emission/absorption relationship of the radiation model, together with its atmosphere temperature and water concentration profile – this is very complex modeling. The Simple Model does have a reasonable theoretical background as it was based on the MODTRAN4 analysis, which should have taken all this into account.

Another possibility is the water produces more clouds. Clouds are very difficult to model, as they are involved in a number of effects. Some clouds absorb incoming radiation, leading to a cooling of the planet. In addition they reflect outgoing IR, leading to warming, but on balance most authors believe there is usually a cooling. Some types of high clouds have little effect on the incoming radiation, but do absorb the IR, increasing the warming effect. Generally, it is assumed that more water produces more cloud which produces a cooling effect from insolation reduction, but this might not be the way the GCM models work.

The magnitude of the effect of water vapor on planet warming is far from clear.

7.1c) Ocean Temperatures

The GCMs perform full heat balancing at each time step, and the surface sea temperature is determined by a balance of heat flows from solar and IR radiation, evaporation, and heat conduction into the body of the sea. Since the sea is vast it takes thousands of years for the small imbalance due to global warming to warm up the whole of the sea.

By choosing one set of parameters for the model simulation – eg high heat conduction from the ocean surface to bulk, for instance, would result in the sea absorbing much of the heat involved in the global warming. This does not alter the final global warming figure, but it does affect the timing to get there (Wigley, 1985). This scenario would suggest that the global warming that we have seen is a fraction of that which it will be at equilibrium. The GCM predictions do suggest that the surface sea temperature (SST) hardly rises. The alarmists say that when the sea temperature finally catches up with the land temperature, the global warming will be extreme.

Another set of parameters might suggest that little heat is transferred into the bulk of the sea because the warmer water stays on the surface, and the sea surface follows the mean temperatures assumed by the Simple Model. Which set of parameters have the GCMs used, and what was their justification?

When we look at the predicted distribution of temperatures around the globe as predicted by the Hadley Centre GCM model – figure 7.3 – we see that the expected rise in SST in the oceans is between 0 and 3°C, only a fraction of the expected mean global warming figure. – with corresponding large temperature rises over the continents. This is in basic agreement with the premise that the SST is low because the surface heat is heating the whole of the ocean – there is the temperature lag. Now if we look at the recorded SST temperature rises so far observed shown in Figure 7.4a, the SST rise is very similar to the mean global temperature rise, as shown in figure 7.4b. suggesting that there is no observed lag between mean global temperature and SST.

It is difficult to explain the differences between the observed and the predicted graphs without concluding the simulated data uses too high heat transfer parameters values. This contradiction is strange, as both sets of data come from the same Hadley Research Centre. This sort of analysis is worthy of deeper study, as data must be available. Without much more detail it is not possible to draw definite conclusions, but one is left with the general feeling that the models transfer too much heat to the ocean depth, – the parameters used may be wrong.

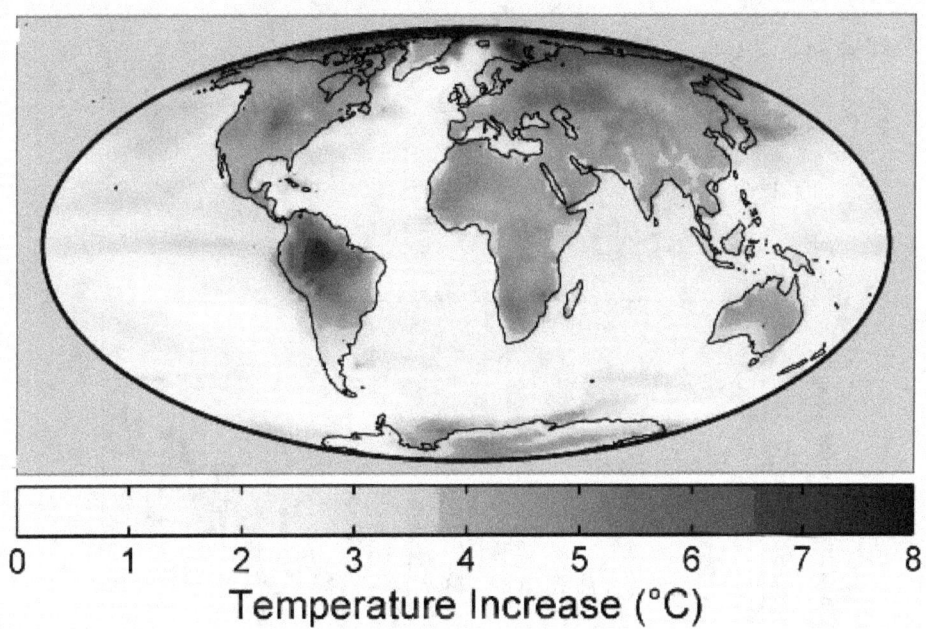

Temperature Increase (°C)

Figure 7.3 Predicted future Temperature Distributions
Since the GCM models calculate the change in conditions segment–by–segment, the results of their global warming studies are expressed as temperature distributions over the planet. The models generally agree that there will be much greater warming over the land masses. The models predict the warming of the oceans will be relatively little.

It is a very serious matter if our observed global warming is only an initial indication of the real effect due to this thermal lag. Against this argument it can be pointed out that there are still no theoretical grounds for explaining the final high global warming, and why is the lag in SST not already observable?

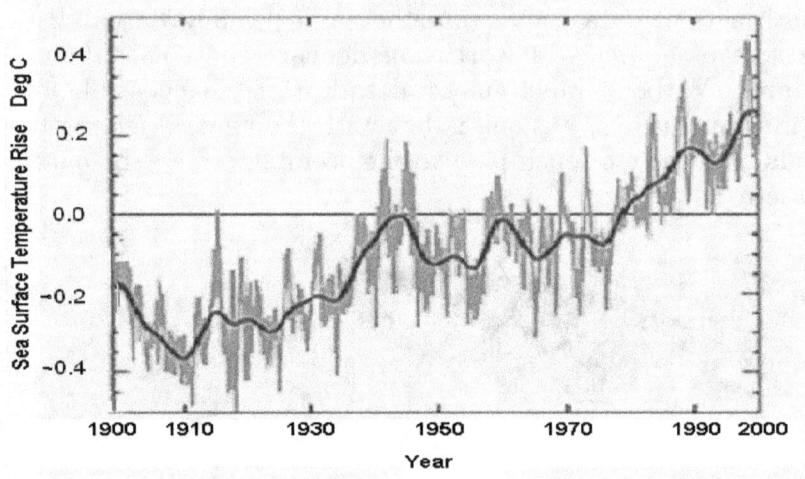

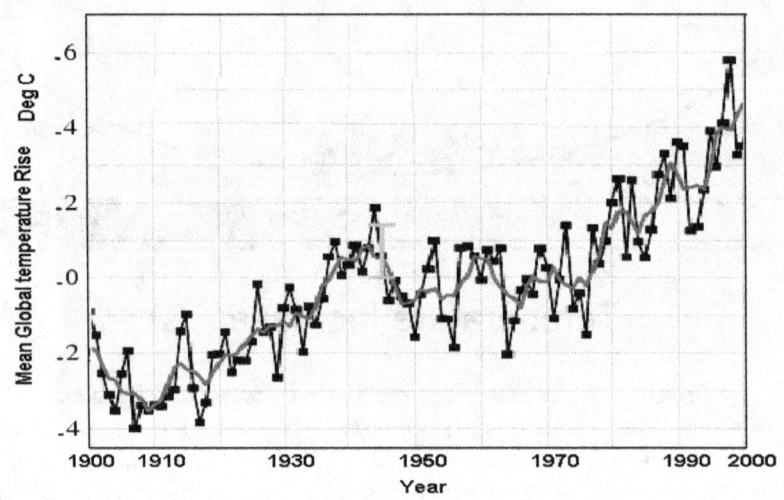

Figure 7.4a and 7.4b Comparisons of Global Warming over Sea and Land

These two curves are remarkably similar, and their Running Means superimpose on each other almost exactly.

Despite the GCM models predicting future warming to be much more severe over the land than the oceans, past records of sea surface temperatures are very similar to the mean global warming figures. There is no evidence from such data that the sea warming lags behind or is significantly different from the land warming.

150

7.1d) Treatment of CO₂ Absorption by the Sea

The absorption of CO_2 from the atmosphere in to the sea, including a proper representation of the carbonate chemistry is a complex subject with very little observed data to check predictions. To date, the Simple Model and the GCMs both show the sea to absorb the same amounts of CO_2. This is understandable as they both agree with the CO_2 levels observed in the atmosphere. Model predictions must agree with observation to be credible models. It depends whether the mass transfer descriptions are the same for all models. If not, the models may well diverge in their predictions of the future.

There may be differences as the CO_2 concentration increases in the oceans. Firstly there is the mixing in the oceans, which is difficult to model, which might result in a quicker saturation of the surface with CO_2 than the Simple Models allow – this would lead to future discrepancies between the models. It may also be that a completely different form of absorption by the sea has been assumed by the GCM models. The Simple Model assumes absorption based on engineering mass transfer theory, but other approaches may have been used by the GCMs.

It is very difficult to quantify the response of the oceans to increased CO_2, but, by the Simple Model selecting its parameters by fitting, and by using the same parameters in the future predictions there is an element of error compensation inbuilt. This might not be the case for the GCMs if their parameters are based on purely theoretical considerations, with less emphasis on tuning.

7.1e) Correlation

As a model has a more complex structure, it contains more parameters, all of which require to be given values. In order to ensure the model gives realistic results some parameters are found by adjusting their value so the model agrees with observation. When a number of parameters are to be adjusted, it is not possible to get unique values for each parameter, – any number of ' sets ' of values may predict observation adequately. However, when the model is used for prediction, the different 'sets' of parameters will give different predictions. This is exactly what is seen with the GCMs. The predictions diverge in the future. So GMC's have problems of correlation between parameter values, and this makes predictions imprecise. The only solution to correlation is to reduce the number of parameters by modeling only those effects thought to be most significant – ie move towards the simpler model.

7.2) Feedbacks

Feedbacks can be considered as a non–modeling approach to the understanding of climates. The basic premiss is that any primary factor causing a change may induce other factors which either increase that change further, or decrease it – hence the concept of positive and negative feedbacks.. This is a useful approach in climatology because the climate changes in the past have been considerable but there is no one change that is large enough to explain the events. Hence there is the concept that one small change induces other changes which together produce the full effect. Taking global warming, the concept of feedbacks is that the primary change – the CO_2, producing a global warming of say 1°C, – produces other changes which either add to this temperature rise, or subtract from it.

Since GCMs have become so complex, it is difficult to look at the individual structure of the model and its equations to determine the correlation factors involved. To handle this problem, the concept of feedback can also be used in climate modeling, where the model is run for a number of conditions, and from the results the 'effective ' feedback (sometimes called 'sensitivity') predicted by the model is picked out and becomes a useful factor for discussing results.

The concept of positive feedback is most easily illustrated by going back to sea ice, considered at the beginning of this Chapter. As temperatures rise, that area of sea ice decreases, and the reflectivity – the albedo–of the sea replacing it is less than the ice. More sunlight is absorbed, less reflected, and the temperature rises further.

With positive feedbacks, the equilibrium is attained very slowly, as each rise induces a further rise, the changes asymptote slowly to equilibrium. With negative feedbacks, equilibrium is attained relatively quickly, because any change is opposed by the feedback and equilibrium is reached more quickly.

Negative feedbacks are good news because they stabilize and minimize changes. It has been suggested that since the planet is 4.5 billion years old there must be stabilizing principles involved and these are provided by biological systems (The Gaia Theory (Lovelock 1989)).

Positive feedbacks are bad news as they can create an instability in the system. If a change in radiative forcing produces a temperature rise, ($\Delta T_{forcing}$) and this temperature rise induces a further temperature rise − say 1 °C rise induces a 'feedback' rise of 0.1 °C , then the total rise will be 1.1 °C − except that the feedback rise will now be 0.11°C , so more correctly the rise will be 1.11°C. This becomes very important when the feed back is 1 °C per 1 °C rise in temperature, making 2 °C − this will now induce a 2 °C feedback rise, making the total 3 °C − repeating this shows that the temperature will rise for ever, there will be no equilibrium temperature (ΔT_{eq}) at which it settles.

Put mathematically, a temperature change, $\Delta T_{forcing}$, induces a feedback temperature change of $f_1 \Delta T_{eq}$. ΔT_{eq}, the equilibrium temperature that will finally be achieved, can be written as an equation:

$$\Delta T_{eq} = \Delta T_{forcing} + f_1 \Delta T_{eq}$$

Which can be solved to calculate the final equilibrium temperature:

$$\Delta T_{eq} = \frac{\Delta T_{forcing}}{(1 - f_1)}$$

or, for multiple feedbacks

$$\Delta T_{eq} = \Delta T_f + f_1 \Delta T_{eq} + f_2 \Delta T_{eq} + f_3 \Delta T_{eq} ..... + f_n \Delta T_{eq} = \Delta T_f + \Delta T_{eq} \Sigma f_n$$

and so
$$\Delta T_{eq} = \frac{\Delta T_f}{(1 - \Sigma f_n)} \qquad (7.1)$$

This equation states that positive feedbacks increase the final equilibrium temperature ΔT_{eq} beyond the original cause of $\Delta T_{forcing}$. Furthermore, when the sum of the feedbacks reaches 1 ($\Sigma f_n = 1$) equilibrium will never be attained. The mathematics of a global model with high positive feedback could show only two stable solutions: an arid dry globe, or a snowball spinning in space.

Feedbacks are the center of the discussion on global warming. Many models predict feedback situations which considerably augment the temperature rise due to radiative forcing caused by atmospheric composition changes. The values of these feedbacks are dependent on the parameters chosen for the feedback equations, which are part of the GCM code. Hence different models give different degrees of warming (ΔT_{eq}) depending on the parameters chosen.

Radiative modeling studies generally agree that a doubling of the atmospheric CO_2, ignoring feedbacks, gives a $\Delta T_{forcing}$ of only 1.0°C, but GCMs models predict different degrees of feedback and this can produce ΔT_{eq} of over 5 °C. A temperature rise of this magnitude would be catastrophic for the planet.

Potential feedback candidates are:

    a) Ice cap areas
    b) Water vapor and clouds
    c) CH_4 release from permafrost and ice melt
    d) CO_2 emissions from warming oceans
    e) Aerosols
    f) Biosystem changes

We will now look at each of these in detail.

7.2a) Ice Cap Areas

The shrinking ice cap areas will induce an increased heat absorption and magnify any temperature rise that is causing the ice to melt. This is well represented in GCMs and results in the common conclusion that the Arctic area will be affected by global warming with temperature increases double that of the other parts of the planet. Figure 7.2 shows that the amount of icecap receiving radiation is very small, and will only get smaller, so there is a common sense feeling that this cannot be a significant reason for warming. Furthermore the resulting heat may stay in the Arctic region, lifting Arctic temperatures rather than affecting other temperatures on the planet.

With a little geometry, calculating the areas of segments of a circle, one can determine the area of ice in 1979 and in 2007. From these areas the mean global radiation change can be assessed and the change in mean global temperature calculated. These calculations are made in the Appendix to this chapter. This analysis suggests that because of the low angle of the sun, the icecap changes do not have much effect on global warming. In fact since 1979 there has been a global warming of about 0.7° C and this has reduced the ice by 30%. The Appendix calculates that the warming due to this melting has been less than 0.15 °C. This gives a feedback (f_{ice}) for the ice cap of 0.15/0.7 = 0.2.

Note that as there is now less ice at the Poles, this effect must reduce, as the maximum further feedback rise due to polar ice when all the ice goes can only be a further 0.11° C. It is therefore unlikely that melting ice feedbacks can play any part in future global warming.

7.2b) Water Vapor and Clouds Feedback

As already discussed, there is the suggestion that there is a considerable contribution made by water vapor to global warming, and this is a positive feedback initiated by the warming due to CO_2. This effect is predicted by some models to be as important as the CO_2 warming, but the reasons for this are not clear.

Chapter 2 discusses the radiative forcing for increasing water vapor content as temperatures increase, and shows that for a 1°C rise in temperature, the 6% increase in vapor pressure induces a forcing equivalent to a 0.29 °C rise. This means a feedback for water (f_{H2O}) of 0.29 can be justified on radiative forcing grounds. Higher feedbacks must include other arguments such as cloud effects.

Clouds are the real unknown in the situation. They are difficult to model, and they may be the reason for the high water vapor feedback claimed by some climatologists.

Water feedback is one of the most unresolved question in the global warming problem. If it is insignificant then global warming is manageable. If it is as high as some suggest then the effect will be catastrophic.

7.2c) CH4 release from Permafrost and Ice Melt

It has been suggested that there are vast quantities of methane locked up in the frozen permafrost in the Northern hemisphere, and as the arctic warms up this will be released, causing more global warming, because methane is a very strong GHG. Likewise, it is thought that there may be much methane locked up as methane hydrate associated with Northern sea ice. Global warming would release this methane as temperatures rise – a classic case of positive feedback.

Against this suggestion is the observation of the CH4 content of the atmosphere over the latter half of the 20th century, where there was measurable global warming, and considerable retraction of the Arctic ice. During this time the CH4 in the atmosphere has risen but has now stabilized see figure 7.5a . There have been no catastrophic temperature rises nor huge releases of methane. For completeness, the methane levels for the last 20,000 years are given in figure 7.5b.

The amount of CH4 released in this period has been taken into account by the analysis with the Simple Model. It is not considered, from the observations so far, that this source of GHG will be significant.

156

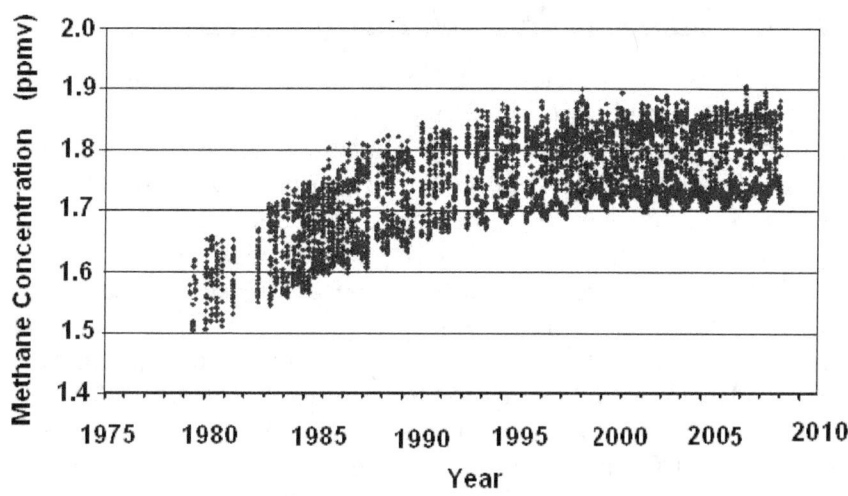

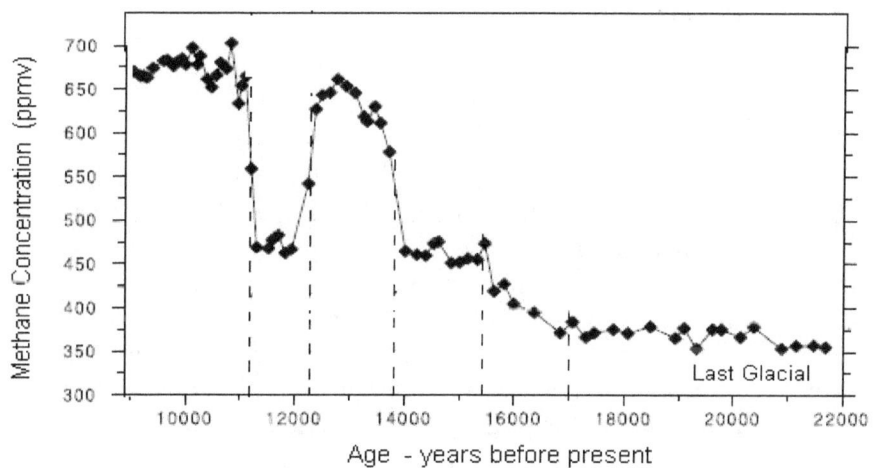

Figure 7.5a and 7.5b Atmospheric CH₄ Concentration Trends:
Recent and past.

After a small increase in the 1980, the CH₄ concentrations appear to have stabilized. Methane levels since the last ice age have been less than half present values, but at no time have methane concentrations been high enough to cause serious global warming. This does not support the suggestion that global warming will cause a catastrophic release of CH₄ from permafrost and methane hydride as the Arctic areas increase in temperature.

7.2d) CO_2 Emission from Warming Oceans

As the oceans warm the solubility of CO_2 decreases and CO_2 from the sea is liberated into the atmosphere which will cause global warming and higher temperatures. Since the sea contains 50 times as much CO_2 as the atmosphere there is ample CO_2 available for the change to be significant. This has been considered a cause of the ice ages, where a change in temperature changes the CO_2 in the atmosphere, which changes the temperature further, so causing the ice age phenomena with its associated changes in CO_2. However calculations show that the effect is not sufficient to be a significant factor in Ice Age temperature fluctuations. In our CO_2 global warming situation, this feedback is inoperative because the anthropogenic CO_2 increases atmospheric CO_2 concentrations well beyond the equilibrium values of the oceans – there is no CO_2 liberated from the oceans as a result of global warming due to CO_2 increases – the opposite is occurring.

7.2e) Aerosols

Aerosols are small particles in the atmosphere which reflect sunlight and so have a cooling effect. These particles can come from volcanoes, and it can be shown that after volcanic activity – such as Pinatubo in 1991 or El Chichon in 1983 – then the planet cools for a few years because of the aerosols (volcanic ash and smoke) which remain suspended in the atmosphere for a number of years and this does reflect some sunlight away from the planet. These effects last only a year or two before the planet settles down to its normal temperatures as the aerosols fall to the ground.

Aerosols can also be man–made. Industrial regions are often covered in a haze which is an aerosol of particulate matter created usually by burning fuels. Aerosols are very difficult to quantify, compared to pure gases which are fully cataloged by spectroscopic analysis. Hence some researchers take the aerosol content of the atmosphere to be the cause of any imbalance in their radiation calculations. Some authors consider aerosols to be very significant coolers of the planet, which means they are effectively masking the full effects on the changes due to GHGs.(Hanson, 2007)

158

The argument that aerosols are hiding the real damage being done so far does require there to be the presence of aerosols permanently in the atmosphere, which is difficult to imagine on a global scale. Though there may be pockets of anthropogenic haze in some areas, the majority of the planet does not seem to be dominated by industrial haze which, once abated, will release a large rise in global warming. Such claims cannot be supported by evidence.

7.2f) Biosystem Changes

The Earth is a planet with its own biosystem. This biosystem has dominated the development of the planet, as living matter has ensured that our atmosphere contains oxygen, much of our land is calcium carbonate, and the sea and the land have attained an equilibrium, in which all matter moves in perfectly balanced cycles. Many of the changes that the planet have undergone in the past have centered on the living systems on the planet. So it should not be surprising if our biosystem responds to global warming, possibly as a negative feedback reducing change. This suggestion was made by Lovelock in his Gaia Hypothesis, and though not discussed in any of our deliberations on global warming, it is not beyond imagination that either the land or sea biosystems will be involved in the changes on the planet with more CO_2.

7.2g) Verification of Feedback

Experimental verification of feedbacks is the best way of revealing their importance, but there is very little published work available doing this. One notable piece of work is that of Lindzen (2009), an MIT climatologist, who has tried to isolate the water feedback problem by investigation outgoing radiation measured by satellite and relating this to the sea water temperatures. By taking an area over the Pacific ocean, he related the outward radiation values to the ocean temperature. He selected periods of time when temperatures either rose or fell consistently, and determined the radiation changes for each of these periods. He also used a series of GCMs to predict the outward radiation. The models all showed, to different degrees, that the raising water temperature reduced the TOA IR emissions, as this is the effect of a water feedback on the temperature rise of the sea. All models predicted a positive feedback, whereas the actual IR measurements displayed a very slight negative feedback. Admittedly Lindzen has spent his lifetime opposing the extreme claims of the global warming and modeling fraternity and it may be that he is biased, but he could well be right and he does understand the need of independent verification of these predicted feedback claims.

Hanson has spent a lifetime warning of the dangers of global warming. He works with a whole range of potential problems. He takes aerosols very seriously, as being an effective cooler at present, hiding the true magnitude of the global warming problem which are to be expected because of feedbacks. He also warns of the heat capacity of the oceans which produces very long time lags in observing global warming, once more claiming that the real magnitude of the problem is being hidden because of the long time before sea temperatures properly reflect the real equilibrium situation. He is a firm believer in feedbacks and makes a good case to declare that present observed data is only a hint at the magnitude of the real problem.

7.3) Summarizing GCM Models

There is no guarantee that the more complete the model is in its scientific description, the more accurate will be the results of any prediction. In the author's experience as a consultant in modeling chemical reaction systems, it was always found necessary to limit the complexity of the system described by the chemists into a 'reduced' model containing only the major characteristics of the system. This was normally opposed by the more scientific brethren as no longer being a scientifically accurate description, but it did reduce the modeling to be within practical limits and was always shown to be effective in predicting those quantities for which the modeling was devised.

The more complex the model, the more difficult to interpret the results, and the easier it is to fit to measured data because of the large number of parameters available to tune to achieve the fit. Predictions then became very questionable. Correlation between parameters compensate for differences within the measured region, but outside the measured region the compensations no longer neutralize each other and predictions become useless. GCM models have been developed for weather forecasting, and include as many features as possible to be scientifically based and are supported by vast quantities of measured data – all past weather records. They have not been developed specifically to determine the effect of increased radiative forcing due to GHGs, where little observational data is available. It may well be that simpler models are more useful for this problem.

7.4) Conclusion

It looks as though we must change our emphasis on the use of the GCMs as the main guide to our future climate. They have hit the problem of correlation, and there appears no way of being certain which of the wide range of predictions is likely to be right. No doubt much more work will be done with them, the coding will become more sophisticated, the computers bigger, and codings will be traded amongst modelers, until no one is sure what is happening, but results will converge as the models become even more similar. But will any of this resolve the correlation problem? Looking carefully at all the factors involved, it is easy to understand the lower predicted values of global warming, but there is difficulty in finding scientific arguments for claiming the higher global warmings that most GCM models predict. It is therefore up to the GCM modelers to explain why their models produce such temperature increases.

However good our science, it still seems difficult to properly define and understand the mechanism and magnitude of feedbacks. There are many climate scientists who believe there will be serious consequences as a result of feedbacks. Their opinions are valuable because of their experience, they should not be ignored. Climate Science is a difficult subject, many factors are far from clear and we must look elsewhere for other clues for us to understand our future. Looking into the past is one possibility. There have been past changes in global temperatures and CO_2 levels. Do these throw light on what might happen? Palaeontology is the subject, and will be the concern of the next chapter.

Chapter 7 Appendix

Calculating the changes in heat absorption with the melting Arctic Ice Cap

A little geometry, calculating the areas of segments of a circle, can determine the area of ice in 1979 and in 2007. From these areas the mean global radiation change can be assessed and the change in mean global temperature calculated.

Since the Earth is tilted with respect to the Sun's axis, we have the 4 seasons and sometimes the poles receive sunlight, and at other times none. To estimate the amount of sunlight received by the ice caps, and how the reductions in the ice caps will produce warming, let us assume there are distinct 4 stages, and calculate the conditions for each season before averaging to produce a yearly figure. Figure 7.6 below summaries these 4 seasons.

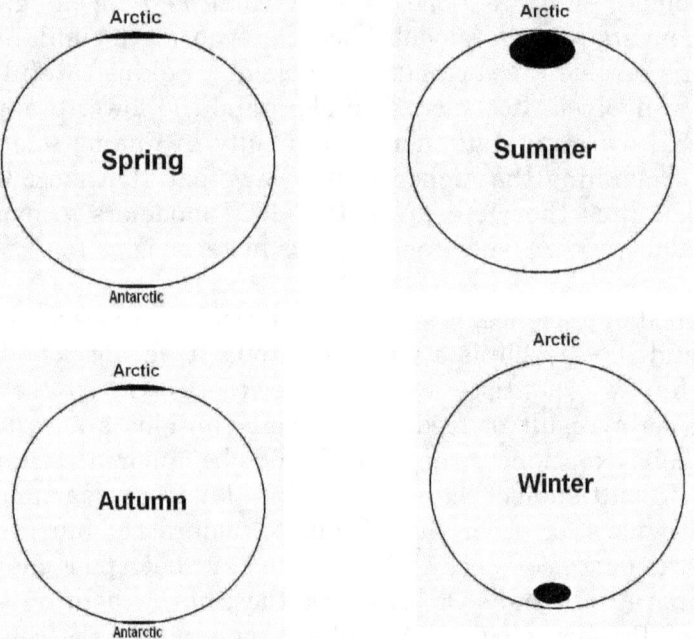

Figure 7.6 Pole exposure to sunlight in the 4 seasons

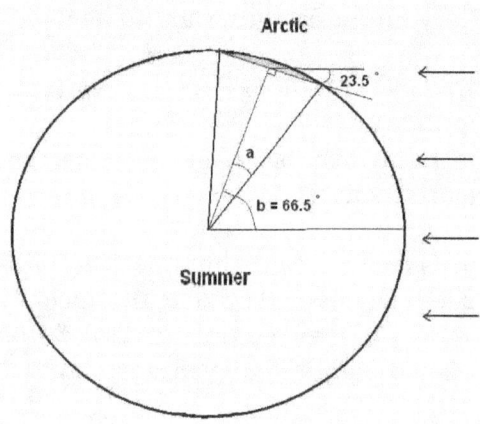

Figure 7.7 Geometric Construction for the Summer Equinox

For Spring and Autumn the analysis takes the form of determining the area of the chord shown on figure 7.2, as follows:

Based on half the angle subtended by the ice cap α:–

Area of iced segment subtended by 2α

$$= \pi/4 \, d^2 \, (2\alpha/360) - (d^2/4) \, (\cos \alpha) \times \sin(\alpha)$$

the total incident area of the globe is $= \pi/4 \times d^2$

so the fraction of global incident radiation involved is

$$Fr_{spring} = Fr_{autumn} = (2\alpha/360) - (\cos (\alpha) \times \sin(\alpha)/\pi) \qquad (7.2)$$

where $\alpha = 90 - (\text{Latitude of extent of ice})$ degrees

For Summer (Figure 7.7) , the analysis develops the equations:

Fraction of global incident radiation (max summer) involved is

$$Fr_{summer} = \sin^2(\alpha) \times \sin(23.5) \qquad (7.3)$$

Chapter 7   Comparing the Models

Taking the mean annual fraction radiated as:

$$0.25 \ Fr_{spring} \quad + \quad 0.25 \ Fr_{autumn} \quad + \quad 0.25 \ Fr_{summer}$$

(Note that this is an overestimate, as it assumes the maximum summer exposure occurs for the whole three months).

Using this expression we find that the fractional area covered in ice in 1979 (α=12deg) involves 0.53% of the incoming radiation. The ice cap in 2007 (α=8deg), involved only 0.22% of global radiation, and the difference in area during the 30 years is (0.53 − 0.22) = 0.31%.

With an albedo change of 0.69 (snow to water), the change in absorbed radiation is

$$0.69 \ x \ 0,31 = 0.21\%$$

Using Stefan's fourth power relationship, as described in Chapter 2, enables us to calculate the change in mean global temperature of 288K due to this change in radiation:−

$$\text{new mean temperature} \ = \ 288 \ (1.0021)^{1/4} \ = \ 288.15 \ K,$$

that is, a mean a temperature rise of 0.15°C .

So the melting ice cap over the last 30 years has contributed a maximum of 20% to the global warming over this time. Furthermore, the contribution to global warming, if the remaining Arctic ice all disappears and the last 0.22% of the ice changes the albedo by 0.69, can only be 0.11 °C.

References

Hansen, J., Sato, M., Kharecha, P., Russell, G., Lea, D., and Siddall, M. (2007), *Climate change and trace gases*. Phil.Trans. Royal. Soc. *A*, 365:1925–1954.

Linacre, E., Geerts, B., *Climates and Weather Explained*, Routledge,1997

Lindzen, R. S., Choi, Y., *On the determination if climate feedbacks from ERBE data* , Geophys . Res. Lett., 36, L16705, 2009

Lindzen, R. S.,The Climate Science Isn't Settled , Wall Street Journal, opinion, 1.Dec 2009
http://online.wsj.com/article/SB10001424052748703939404574567423917025400.html

Wigley, T. M. L., Schlesinger, M. E., *Analytical solution for the effect of CO2 on Global mean temperature,* Nature , 315, 649, 1985

Further Reading

Gaia, *James Lovelock's The Ages of Gaia* , 1989
see http://erg.ucd.ie/arupa/references/gaia.html

G W Paltridge, *The Climate Caper*, Connon Court publishing, 2009

Washington, W. M. and Parkinson, C. L., (2005), *Introduction to 3−Dimensional Modeling,* University Science Books, US

Chapter 7 Comparing the Models

Chapter 8 Looking at Palaeontology for Evidence of Climate Sensitivity

Because the different GCM models of the climate give no clear conclusions on the magnitude of likely global warming – they just present a wide range of possibilities – we must look to other, separate evidence to investigate the importance of CO_2.

In the distant past there has been much more CO_2 in the atmosphere, and CO_2 levels have oscillated considerably during the life of the planet. Hence, by studying past climates – palaeoclimatology – there may be clues as to the relationship between CO_2 levels and global temperature, and we may also learn what the long–term effects of having higher levels of CO_2 are likely to be.

The planet has gone through endless oscillations in temperature, which have resulted in cycles of ice ages. The atmosphere has contained different levels of CO_2, identified by air bubbles trapped in antarctic ice, where the past time is captured in deep glaciers.

Important questions to ask are;

What is the past concentration history of CO_2 on the planet?

Is there a cause/effect relationship between CO_2 and global temperature?

Did warming precede the CO_2 increases or did CO_2 increase precede the warming?

What is the history of atmospheric CO2 levels, and the DIC, pH and alkalinity in the sea ?

Have there been changes in biological activity in the sea at the different times?

All these questions are very relevant to our discussion of the future conditions on our planet, but how can such past information be collected? Luckily, the world is very complex, and out of this complexity we are able to tease out a surprising amount of information.

Chapter 1 explained that all atoms of any element are not identical because of isotopes. Studying isotopes gives us an entry into looking at the past. Since this book is attempting a critical look at all the scientific arguments, we should at least have an appreciation of how the evidence was deduced. This means going into a little understood subject – isotope analysis.

8.1) Isotope Analysis

Although the number of protons and electrons in an atom define which element it is, the number of neutrons in the atomic nucleus is not fixed. The number of neutrons change the mass of the atom, but not its chemistry or combining properties. So when a compound reacts or changes physically by diffusion or evaporation, the *rate* of these changes, which is affected by the mass of the molecule, will be different for the different isotopes.. The lighter isotopes are generally faster, and so when changes are only partial, the lighter isotope concentration changes most, and this difference in isotope concentration can be detected by mass spectroscopy. This is the basis of the science of *Stable Isotope Analysis.*

The isotopic composition of any substance may give a clue to its past history. There are many isotopes and very many different situations where the isotopes behave differently, and so leave clues of their past. There is no simple formula to know where isotope analysis can be useful. It is pure detective work in looking for effects which have different isotopic rates which may give useful traces as to how the material has got from where it was to where it is now.

Analysis of isotopes is done by a technique called mass spectroscopy, where the material is converted into a gaseous form, ionized and the ionized beam bent in a magnetic field. The degree of bending depends on the mass. Hence the target of the beam receives a spectrum of lines because of the different amounts of bending depending on the different mass of the particles. The position in the spectrum gives the mass, and the height (signal magnitude) defines its amount. The method is looking for ratios of different isotopes, not absolute quantities, and so there is no need to use an accurate measured quantity of sample.

The Appendix to this Chapter goes into more detail on the definitions and applications of isotope analysis for those reader wanting more information.

Four examples of application to palaeontology are now illustrated.

Example 1–^{13}C isotope analysis for identifying sources of CO_2

Plants take up carbon from the atmosphere with a slight preference for the lighter isotope. Hence different forms of plant life have different quantities of the heavier isotope ^{13}C – as shown by figure 8.1. This means that we know that fossil fuels have come via plant material. We also know that as the amount of lighter isotope in the atmosphere has been steadily increasing over the last century, this means the natural atmospheric CO_2 is being mixed with CO_2 originating from plants. This is evidence that CO_2 from fossil fuels is now significant in the atmosphere, as is discussed in Chapter 1.

Example 2 ^{18}O isotope Oxygen for determining Temperature

In the carbonate equilibrium in seawater, the equilibrium is slightly different for CO_2 containing the heavier isotope of oxygen compared to the equilibrium with the normal isotope, and this difference is dependent on temperature . This is shown on figure 8.2. The deposited skeletal material from the past is made from carbonate, and so the heavy ^{18}O isotope fraction of that carbonate is dependent on the temperature of the water at the time of its formation. In this way we know more about past temperatures.

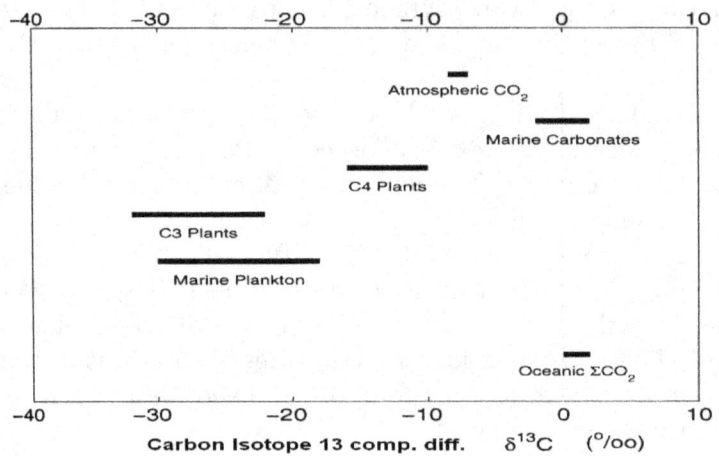

Figure 8.1   Distribution of ^{13}C isotope

The ^{13}C isotope distributes itself differently in plants because it is less preferentially captured in photosynthesis. Hence fossil fuels have a lower ^{13}C than normal. The reduction in ^{13}C in the atmosphere over the last 100 years is evidence that the normal CO2 is being mixed with CO2 originating from plant material.

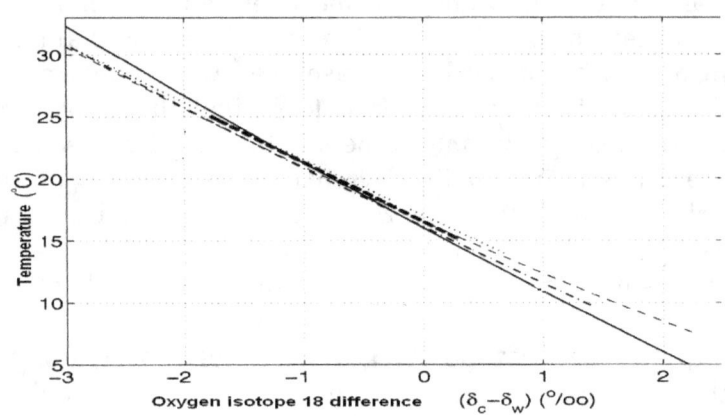

Figure 8.2 The Relation between 18O and Temperature

Different researchers all agree that the oxygen isotope composition of carbonate material is dependent on the temperature at which the carbonate was formed, because the carbonate/bicarbonate equilibrium differentiates between oxygen isotopes. Hence the sea temperature at the time of the formation of precipitated CaCO3 can be determined by ^{18}O isotope analysis.

Example 3 Boron ^{11}B isotope for determining pH

Boron in seawater is present in equilibrium between 2 states – one 3 valency and one 4 valency. This equilibrium is controlled by the pH of the water but the equilibrium is slightly different for the two isotopes of boron. So by analyzing the ratio of the boron isotope concentrations of precipitated material, it is possible to conclude the pH of the medium from which the material was precipitated. We therefore know something about the past acidity of seawater.

Example 4   Oxygen isotope ^{18}O for determining sea levels

Water containing the heavier isotope of oxygen has a lower vapor pressure than water with normal oxygen. This means that when water evaporates, the vaporized water (rain and snow) contains less of the heavy oxygen isotope and conversely the sea contains more. The water falls as snow and, during an ice age, stays for thousands of years increasing snow coverage at the poles. The sea water level falls, and the heavy isotope of oxygen in the sea increases, and this is reflected in the isotope levels in deposited materials. The change in heavy oxygen isotope concentration leads us to be able to calculate the resulting sea level. Figure 8.3 shows the range of oxygen isotopes for different components of the planet surface, and figure 8.4 shows a prediction of past sea levels using this method.

8.2) Paleontological Dating

It is very important to be able to give dates to the information obtained from bore samples of sediments and icecaps. The usual method is that of ^{14}C carbon isotope dating, where the quantity of the radioactive isotope of carbon, ^{14}C, is measured, and this enables its date of fixation to be set.

Due to cosmic proton (p) activity in the upper atmosphere, radioactive ^{14}C is continuously being formed from nitrogen

$$^{13}N + p = {}^{14}C$$

The ^{14}C decays radioactively, so there is an equilibrium standing concentration set up with atmospheric CO_2 remaining at a standing

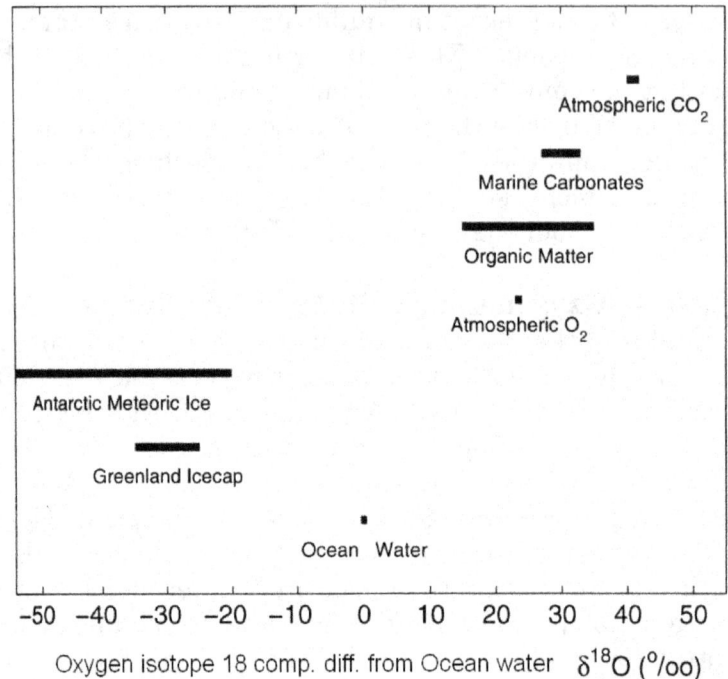

Figure 8.3 Variations in ^{18}O content of some Terrestrial Compounds

Water containing the heavier ^{18}O isotope has a lower vapor pressure than the normal ^{16}O containing water, and so water vapor, rain and snow are depleted in ^{18}O. This depletion increases the ^{18}O content of the sea, and so from this change the quantity of ice at the poles can be calculated. By subtracting this quantity of ice, the resulting sea level can be estimated.

$^{14}CO_2$ level. When CO_2 is removed from the system, this equilibrium is no longer maintained, and the ^{14}C decays, isolated from exchange with the atmosphere. With a half life of 5000 years, its decay can accurately predict ages from a few centuries up to 60,000 years BP (Before the Present). For times longer than 60,000 years, other radioactive methods are available, but based on other isotopes originating at equilibrium in the magma of the earth's mantle.

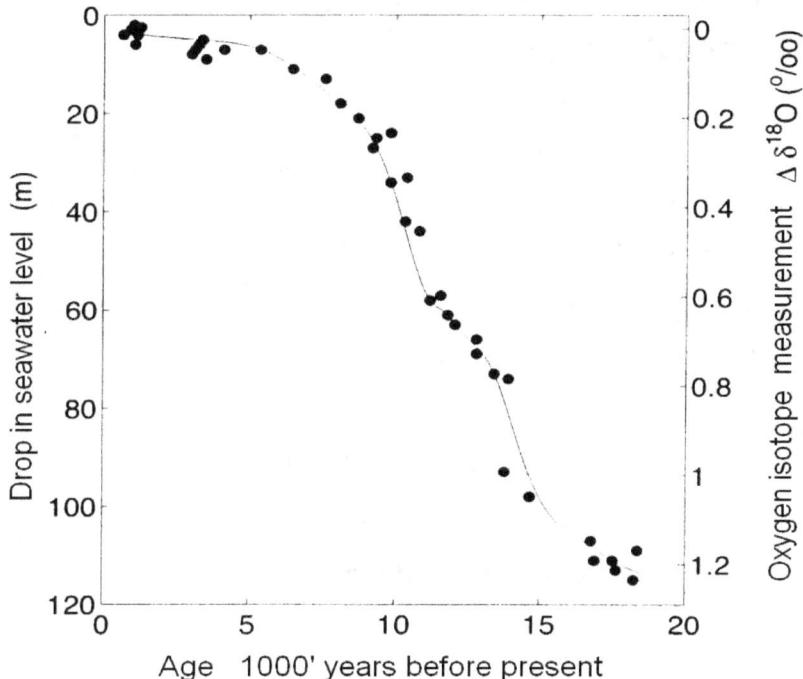

Figure 8.4 Sea Levels since last Ice Age
As determined by ^{18}O content, from analysis of coral samples . The drop in ^{18}O content reflects the melting of the ice caps, which enables the changes in sea level to be calculated.

Once these are fixed in solidified rocks, the various radioactive isotopes then start to decay, and this gives an indication of the age of the rock formation. Choosing the right isotope for the investigation enables the age of interest to be covered. For isotopes other than carbon, the method can only be applied in association with volcanic events where new rocks are formed or volcanic ash is deposited.

Beyond this there are the stratigraphic methods which rely upon the position of the material in question in relation to other materials which can be dated. Incorporated in stratigraphy is the investigation of fossil details, which are known to be related to specific ages.

8.3) The Relationship between Past Temperatures and Atmospheric CO_2

Data on the Earth's past has been obtained by boring into the ice of the Antarctic. Ice samples contain bubbles of air trapped as the loosely packed fresh snow is compressed to ice, and these bubbles can be analyzed to determine the atmospheric CO_2 level at the time the ice was formed. Isotopic analysis is essential to obtain information from these borings. Dating is done by ^{14}C analysis, sea level by $\delta^{18}O$ analysis and temperature by δD and $\delta^{18}O$ analysis.

We can look at the planet's past temperature and CO_2 history in 4 stages – 1000, 20,000, 500,000 and 500,000,000 years BP.

8.3a) Over the last 1,000 years

Over the last 1000 years the planet temperatures have been reconstructed using a variety of methods and the well-publicized 'Hockey stick curve', figure 8.5, has been the result.

This curve shows that for the last 100 years there has been an upturn, rather like a hockey–stick with a flat handle representing the preceding 1000 years. After some controversy, it was agreed that the 'handle' was not quite so flat, and there was a 'little ice period', in the 1600s, and in the 1400s there was a warm period, approaching today's temperatures.

Overall, the hockey stick curve indicates that there is cause for concern. There has been little change in CO_2 during the last 1000 years until the recent anthropological period. Temperatures have been reasonably stable until the last 100 years. Both these points indicate that there is a new condition emerging in the last hundred years, and it is most likely anthropological.

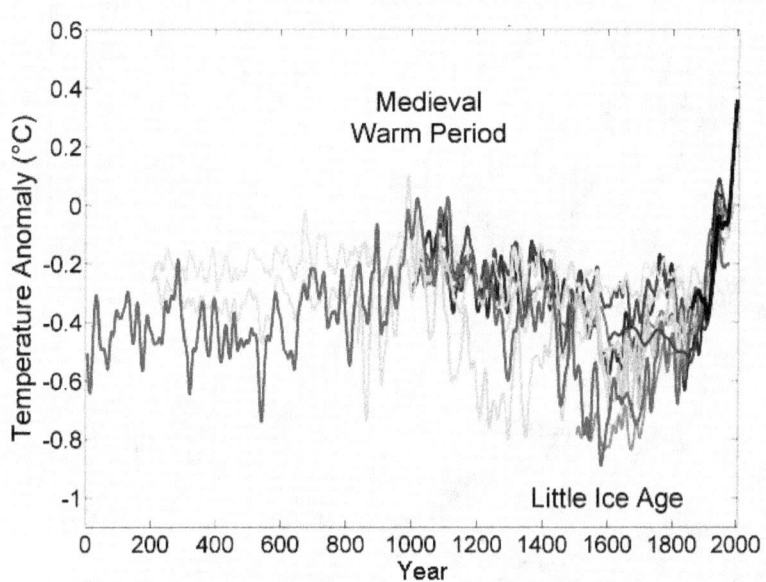

Fig 8.5 The 'hockey stick' reconstructed temperature record for the last millennium

Apart from a Little Ice Age and and a Medieval Warm Period, planet temperature levels have been remarkable constant until the last 100 years. This curve dramatically shows that something has happened since the beginning of the industrial revolution.

8.3b) Over the last 20,000 years

A detailed study (Monnin, 2001) of ice core data over the last Glacial termination – the time since the last age – shows how detailed the information can be from the sea ice borings. Figure 8.6 show the isotope data recorded. There is a remarkable correlation between the temperature δD, and the CO_2 curves. This is very convincing evidence that there must be a relationship between them. Careful analysis of the timings suggests that the temperature rises 800 ± 600 years before the CO_2 rises. This suggests that the CO_2 is not the cause of the temperature rise, but the authors point out that

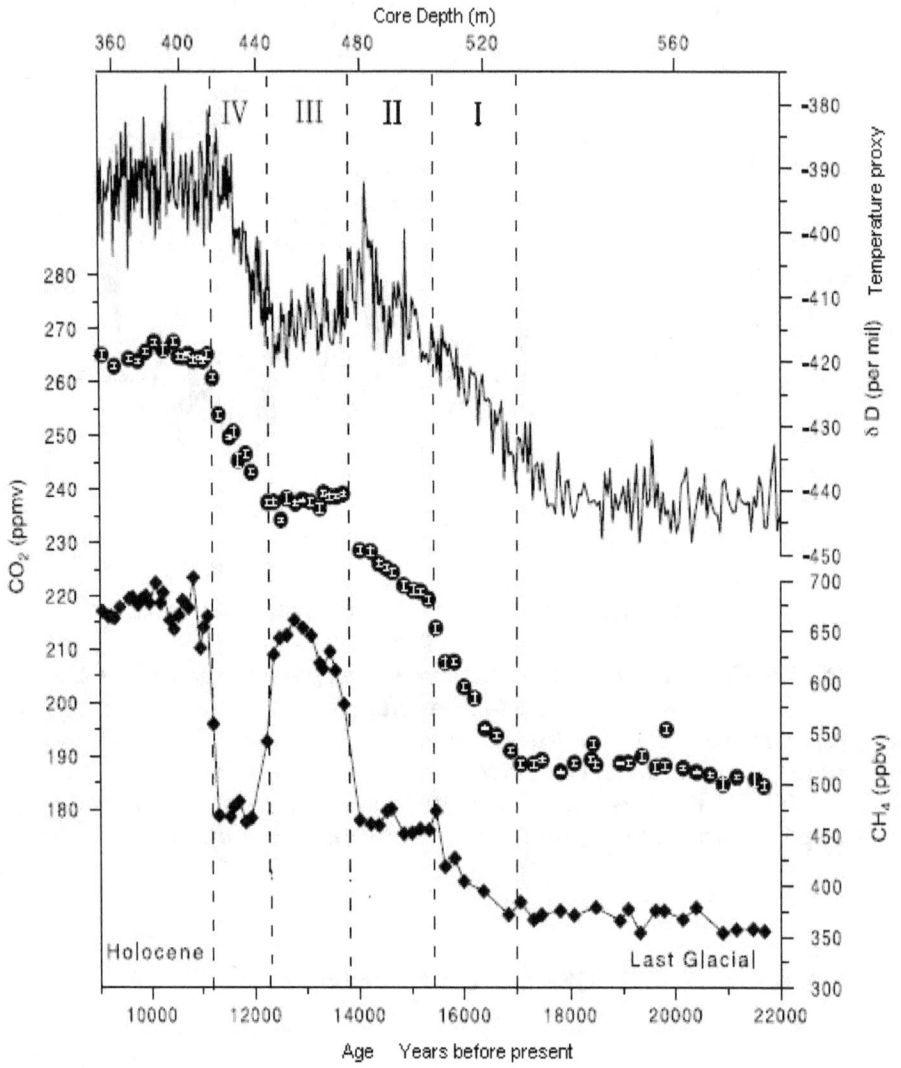

Fig 8.6 20,000 years of ice core borings

Note the remarkable correlation between the temperature, and atmospheric CO2 and methane levels since the last ice age. There must clearly be a cause and effect here. The division into 4 discontinuous regions suggests a change at these interfaces. It is suggested that they are due to discrete changes in ocean circulation patterns. .

176

the times are so close this does not rule out CO_2 being an amplification factor of the temperature rise that is occurring. Interesting also is the trace of the CH_4 concentration, which also correlates with the temperature and CO_2 curves. The CH_4 is thought to come from the Northern Hemisphere wet–lands when the Northern latitudes warm up. This is one of the concerns of today's global warming.

Explanation for the temperature and CO_2 rises center around possible changes in the ocean circulation patterns. The 4 distinct phases marked in fig 8.6 are thought to represent discrete changes in ocean circulation patterns of the Thermohaline and the North Atlantic Deep Water currents, which are associated with the Gulf Stream.

8.3c) Over the last 500,000 years

The Vostok borings from the Antarctic ice sheet have enabled the variation in temperature, determined by isotopic analysis, and CO_2 composition determined by direct analysis, to be recorded as far back as 500,000 years. Vostok has been one of a number of projects, such as the Deep Seas Drilling Project and the Ocean Drilling Program, which have extended the picture of the history of the biology, temperature and composition and level of the sea in the past. Figures 8.7 show some of the temperature measurements over the last 500,000 years. Figure 8.7 does show distinct oscillating temperatures, and with it matching oscillating CO_2 levels

Isotopic analysis also enables us to look at the equivalent sea levels and ice volumes and this also oscillates in phase with the temperatures and CO_2 levels, as shown in Figures 8.8.

This data allows us to investigate the relationship between CO_2 and temperature. Careful research again concludes that the rise in CO_2 occurs 600 ± 400 years after the rise in temperature (Fischer, 1999), suggesting that the CO_2 change is the result of the temperature change, not its cause.

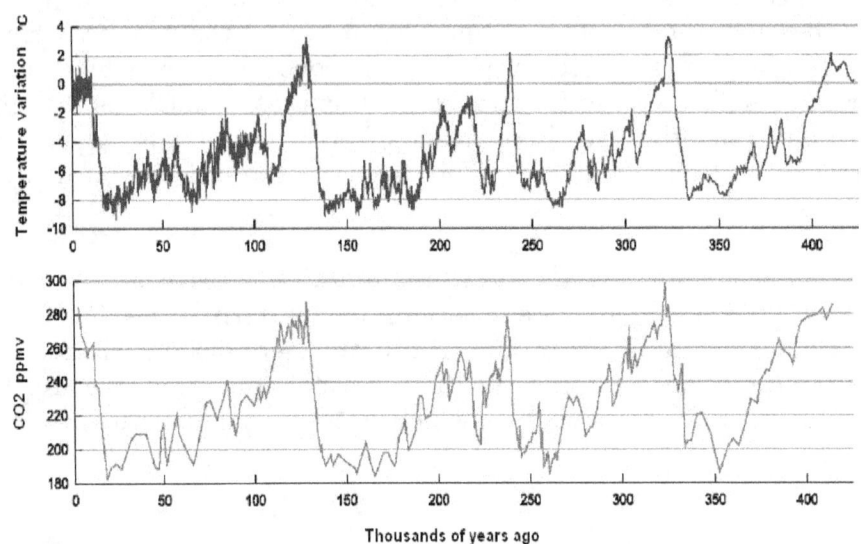

.Figure 8.7 Vostok core sample measurements over 500,000 years
This 500,000 year range shows how much oscillation there is on
the planet. The Milankovic 100,000 year cycles are clear to see.
The 20,000 cycles are rather more difficult. The CO_2
concentrations follow the temperature pattern meticulously.

Why does the CO_2 increase during the warming after the ice age?
Any increase in atmospheric CO_2 long–term implies that the
atmosphere is in equilibrium with a sea with a greater CO_2 partial
pressure. This increase in the partial pressure of the sea cannot be
due to an increase in CO_2 content (DIC) of the sea, because there is
no explanation for where all the extra carbon can have come from. It
must mean that the partial pressure of the sea has changed because
sea pH, alkalinity or temperature has changed.

As the sea warms, the CO_2 partial pressure of the sea increases,
– is this the explanation linking these events? A simple quantitative
analysis, similar to Chapter 3 Appendix, shows that the CO_2
changing from 180 to 280 ppm does not fit with the the few degrees
change in temperature. This therefore cannot be the cause as it can
only explain about 20 ppm increase in CO_2, not the full CO_2
increase of 100 ppm.

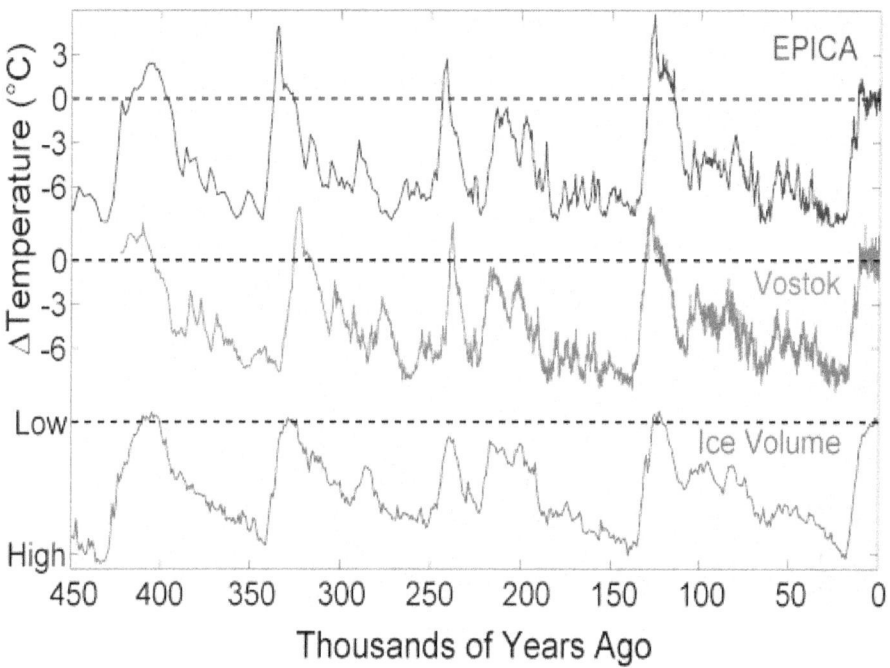

Figure 8.8 core samples over the last 500,000 years, from 2 sources
The EPICA data is a separate project to Vostok, yet the temperature data is in remarkable agreement. This figure also shows the ice volumes, predicted from the ^{18}O concentrations at different levels. Again there is very good agreement between the temperature history and the deposition of ice.

This leaves the explanation to be in the changing seawater composition, which would cause the sea CO_2 vapor pressures to match the observed CO_2 atmospheric compositions. A change in the alkalinity means a change in the carbonate balance of the oceans, which involves a reduction in the Ca^{++} ions in the sea water – which can only occur by means of biological action, building more $CaCO_3$ skeletal debris and precipitating it on the seabed. Other biosystem changes will occur after an ice age, where land becomes free of ice and becomes colonized by biological systems. The sea level rises and open sea area may also give rise to more biological activity.

8.3d) Over the last 500,000,000 years

Over the last half a billion years the temperature change is thought to be as shown by fig 8.9 indicating that the planet temperature has been reasonable stable for all of the time, despite the very high CO_2 concentrations.

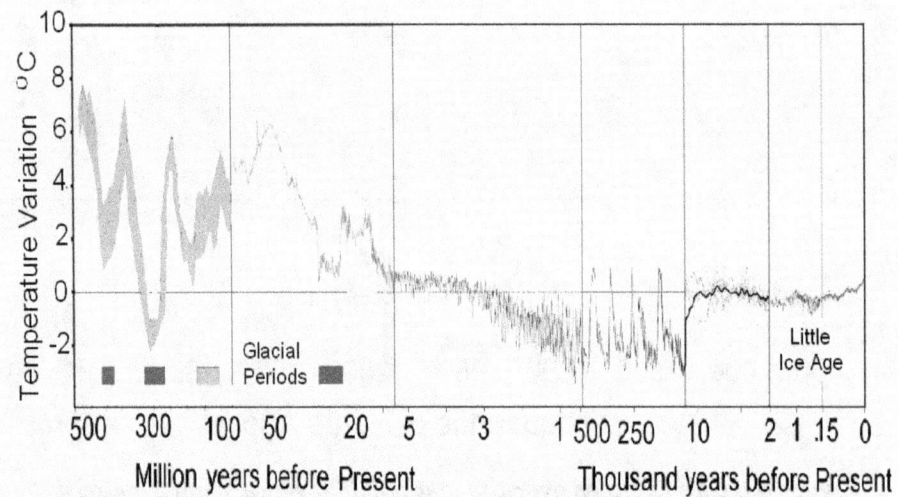

Figure 8.9a Planet temperature estimates over 500 million years
Periods 50 to 500 million years have seen CO_2 levels in the 1000 – 5,000 ppm levels, and over this time planet temperatures have varied between –2 and +6 K about present day temperatures. . This does not suggest there is a strong link between planet temperatures and CO_2 concentrations.

An investigation of the effect of very high concentrations of CO_2 can be made by studying some of our near–neighbor planets. These planets have much higher CO_2 levels, but the planet temperatures can be quite adequately explained without invoking warming due to the high levels of CO_2.

No evidence can be found from such observations of our past that CO_2 is a primary cause for planet warming.

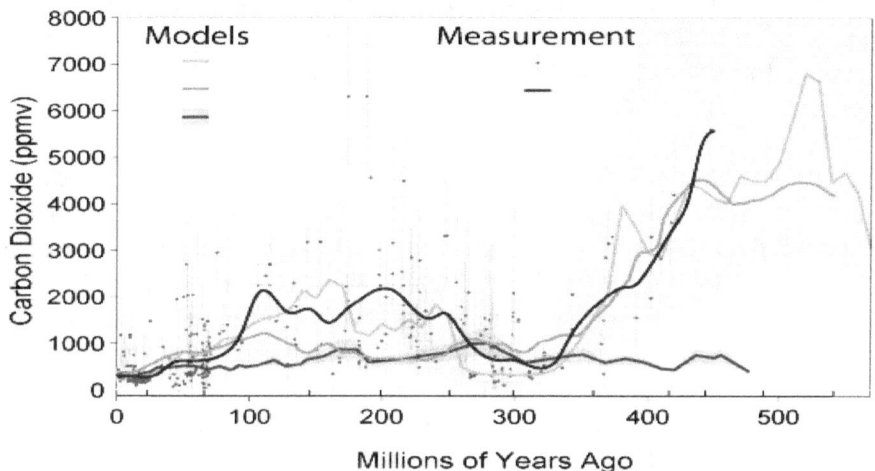

Figure 8.9b Planet atmospheric CO2 estimates over 500 million years

Figure 8.9a and 8.9b Temperatures have remained constant within 2 –3 degrees, for the last 50 million years. Before that time temperature oscillated over a –2 to +8 ° C range, during which time the CO2 concentrations changed to 300 ppm from 1000, – 2000ppm; 400 million years ago saw CO2 levels of 3000 – 5000 ppm; yet temperatures still remained within the 10 ° C range

A review article (Sigman, 2000) gives a thorough summary of the plausible hypotheses that have been presented as explanations for the CO2 pattern associated with ice ages The list below summarizes each of the hypotheses,.

Carbon storage on the land; – the shift of land storage of CO2 to the sea as ice covered much land area could produce CO2 oscillations. This contribution to CO2 changes has been calculated as being small.

CO2 storage in the oceans – the slow circulation of deep ocean water to the surface might result in oscillations in atmospheric CO2. Again, the contribution to CO2 changes is small.

Ocean temperature – as the water temperature increases , CO2 is liberated into the atmosphere due to solubility effects. Again, the contribution to CO2 changes is small.

Ocean carbon cycle and alkalinity changes – weathering of CaCO3 resulting in changes in alkalinity can produce atmospheric CO2 changes, but there needs to be reasons for these carbon cycle changes to occur.

Nutrient changes resulting in different bioactivity – this provides a reason for the changes in the carbon cycle. CO2 is liberated or absorbed from the sea when there are alkalinity changes. Alkalinity changes occur when CaCO3 is deposited in the sea bed or re–dissolved from the seabed, and this involves the biosystems which produces the CaCO3. When there is more growth, alkalinity changes and CO2 change can occur. Explaining why there are changes in the biosystem growth leads on to discussing their growth–limiters such as nitrates and phosphates in the sea. Changes in these nutrients can occur as the result of changes in deep ocean currents which expose the surface waters to more of these nutrients – causing increased growth. Going further with these discussions is not a matter for this text, but it does show how complex Earth Sciences are.

This last hypothesis is considered by the authors of the paper to be the most likely cause of the CO2 patterns that are associated with past ice ages .

8.4) Ice Ages, Planet Sensitivity and Feedback

Throughout the ages there have been periods where the north ice covered much of the northern hemisphere. Figure 8.10a and b show the extent of the ice coverage during an ice age. The cover goes as far as central Europe and mid North America.

Figure 8.11 shows the extent of the sea ice in the northern hemisphere. It can be seen that in the northern hemisphere the ice covered a significant fraction of the globe. In fact from figure 8.11 the coverage can be calculated as being equivalent to reaching as far south as the 48° latitude.

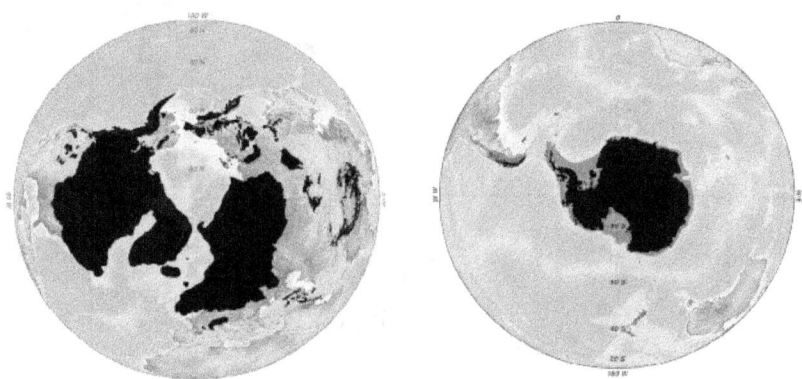

**Figure 8.10a and b   Extent of glacial ice coverage of the continents
at the height of the last ice age**
*The northern hemisphere with its greater land mass was more
affected by the glaciers in the   ice age than the southern
hemisphere. Glaciers crept down covering half of the USA and
down to mid–Europe. Ice depths were 3 – 4 km, and this reduces
sea levels by 120 m.*

It is interesting that it is the southern hemisphere that is not
affected by glaciation. This no doubt is due to the greater land mass
in the north, where fallen snow can build up. With less land and
more water, as is the case for the southern hemisphere, there is
better heat transport by the ocean currants, and less chance of
permanent coverage of snow and eventually ice.

The first clue towards an explanation for ice ages was given by
Milankovic who pointed out that the temperature cycles were
basically every 20,000 and 100,000 years. These periods correspond
to the oscillations of the planet caused by its rotational axes. The ice
age periods corresponded to those periods when the planets received
less radiation from the sun. However this is not the full explanation,
because the radiation differences from the orbital changes are
insufficient to cause temperature changes of that magnitude on the
planet. It appears that the Milankovic cycles may trigger the onset

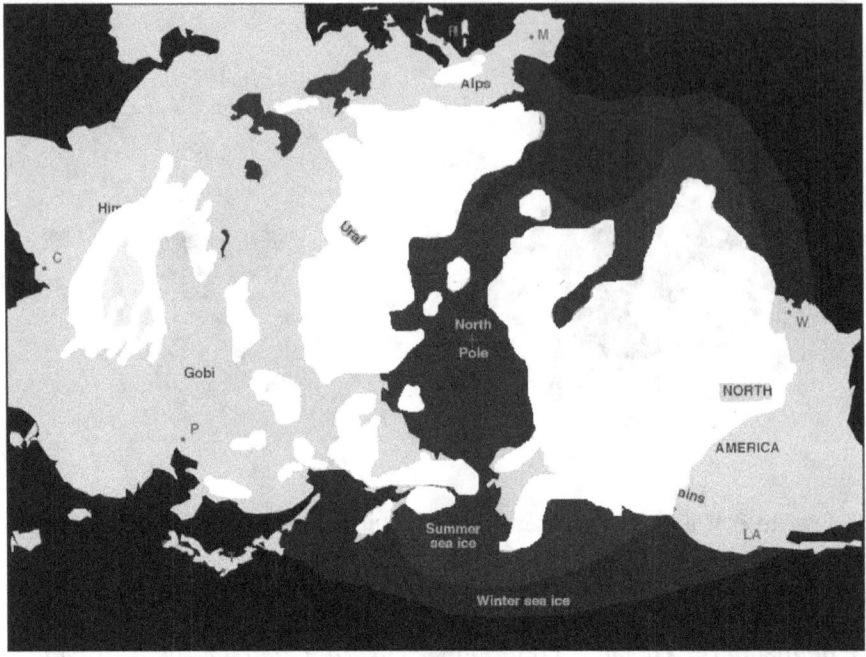

Figure 8.11 Extent of the land ice coverage at the height of the last ice age

The white areas show the extent of the glacial coverage at the last ice age. Winter and summer sea ice was also correspondingly greater. The proportion of the planet covered by ice was considerable. The cause of the lower planet temperatures is attributed to the change in albedo – with more reflection of sunlight, and less absorption by the ice.

of the ice age, but are magnified in some way by other factors. This is the evidence that the climate does have feedbacks which magnify changes, – one of the major worries for the future climate.

The most likely feedback is the change in albedo when the sea or land is covered in ice, This feedback is discussed in the last chapter under sea ice, where it is suggested that the small amount of sea ice, high in the northern latitude is not likely to have any large influence today. This argument does not hold when the ice layer creeps down as far as mid Europe and mid USA, as happens during the ice ages. The albedo effect with so much ice cannot be discounted, and is sited as the most likely feedback which could convert the weak Milankovic cycles into a magnitude which creates ice ages.

Knowing the extent of the ice coverage at the height of the ice age, and the albedo of ice, water and land, we can calculate the reduction in radiation absorbed, and convert that to a mean global temperature change, as was demonstrated in chapter 7, with ice reaching the 48°latitude. The formula in Chapter 7 Appendix calculates that, during the height of the ice age, 7.5% of the earth's radiation impacts ice and snow. This 7.5% now reflects 80% of the light falling on it. replacing the ice–free reflectivity of the northern hemisphere of 30% water and 70% forest, with a mean reflectivity of 11% (Linacre, 1997)

7.5% of the surface which had absorbed fraction $(1 - 0.11) = 0.89$ will now absorb $(1 - 0.8) = 0.20$, fraction of the incoming radiation (S_0) – a decrease of $0.89 - 0.20 = 0.69$ on 7.5% of the globe surface.

The whole globe has a mean albedo of 0.7 and so the decreased absorption due to the ice cap, expressed as a fraction of the total energy received by the planet, will be

$$0.075 \text{ x } S_0 \text{ x } 0.69 / 1.0 \text{ x } S_0 \text{ x } 0.70 = 0.073$$

So the ice cap will have reduced the radiation absorbed by the planet by 7.3%

As in Chapter 7 Appendix, we can calculate the temperature change that this radiation difference will induce as

$$288 \, (1.073^{\frac{1}{4}}) = 293.1 \text{ K, or a temperature difference of } 5.1°C$$

This temperature difference of 5.1 °C will cause a change in the water vapor pressure in the atmosphere, and as shown in Chapter 2, each 1 °C change in temperature will induce further temperature change of 0.29 °C –(ie the feedback (f_{H2O}) for water is 0.29). The CO_2 changes during an ice age from 180 to 280 ppm and this will produce a warming of 0.5 °C. Adding these effects together, using the feedback relationship developed in Chapter 7 gives a total temperature change (ΔT_{equ}) of :–

$$\Delta T_{equ} = (5.1 + 0.5)/(1.-0.29) = 7.9\,°C$$

This is in surprisingly good agreement with temperature cycles of 8°C during the last ice age – see figure 8.7. There is no need to look for other feedbacks beyond the ice coverage, CO_2 and water infrared absorption to explain why the planet can cool as much as it does during an ice age, once triggered by the Milankovic cycles.

After the minimum temperature, the melting ice allows more radiation to be absorbed, there is again a positive feedback from the ice, and this will enhance the recovery from the ice age, giving the 8°C cycle. Exactly why the growth of ice reverses is not known, but it is not difficult to imagine in any oscillation system there are subtle changes that occur which reverse the direction of change – that is why they are oscillation systems.

There are other possible suggestions that have been cited for the cause of ice ages –

> changes in ocean circulation patterns,
> cessation of the gulf stream and hence the cooling of the
> > northern hemisphere
> volcanoes
> continental drifts
> meteorites

The problem with all these explanations is that they cannot be linked to a regular cycle, which is the major characteristic of the pattern of global temperature.

We should report here the claim made by some climatologists that the CO_2 and its greenhouse effect must be the prime mover in ice age temperature changes, and that the large temperature fluctuations are the result of the small 100 ppm CO_2 increase plus the feedbacks that they induce. This relies on the belief that there must be feedbacks involved which we do not fully understand. They argue that we have not seen this temperature rise with the anthropogenic

CO_2 rise yet because it is held back by ocean temperature lags and anthropogenic aerosols. When the full effect of the CO_2 is felt, it will be truly catastrophic for the climate. These views are held by a number of climatologists, the most prominent being Hansen (2007).

This hypothesis is never cited by palaeontologists. As a theory it has a series of weaknesses – it does not explain the regular temperature cycles and it requires there to be feedbacks so far unidentified. If the main feedback is the ice coverage, then this feedback cannot be active in present day conditions, as there is now little ice. Nevertheless, it is a theory firmly held by some climatologists.

8.5) Conclusions

The temperature and CO_2 patterns of the planet in the past can be explained in terms of Milankovic cycles together with feedbacks through ice sheets, water and CO_2.. There is no need to invoke explanations involving CO_2 causing the temperature changes with other feedback effects. There is little support from paleontological data for the suggestion that CO_2 had been the cause of planet temperature instability. In fact the information suggests that the planet temperature levels have remained fairly constant even at very high concentrations of CO_2, as was the case in the early history of the planet.

The CO_2 changes that have occurred in the planet are difficult to explain without invoking variations of alkalinity of the sea through changes in the biosystems and the precipitation and resolution of the $CaCO_3$ in the sea. This indicates that once the CO_2 reaches the sea, this is not the end of the story and it will be assimilated by the sea, changing alkalinity, which will protect the sea's biosystems. What is not known is the speed at which these changes will occur. The paleontological data can be assumed to represent equilibrium conditions as time scales are in thousands of years. For our global warming problem, and any associated changes in the sea require alkalinity changes in matters of decades, rather than thousands of years. The rate at which CO_2 will be involved in changing alkalinity of the sea is not well researched but is very important in our study of the final effects of fossil fuel burning.

Chapter 8 Appendix

Isotope Analysis

Chapter 8 explains that isotope analysis is a very important tool in investigating conditions on the planet in the distant past. This appendix goes into some of the details of these methods. The technique has its own nomenclature. It needs to define deviations from 'normal'. Where 'normal' has firstly to be defined. Deviations are then expressed as δ in parts per thousand (‰).

Once definitions are completed, it is detective work to find what compounds have a range of isotopic concentrations, and to see what they can tell us about the past.

This Appendix goes into the finer detail, and gives examples of where it has been useful

Definitions

Nomenclature

The isotope is written by preceding the elemental formula with its molecular weight: eg the heavy isotope of oxygen with a molecular weight of 18 is written ^{18}O, and the ratio of the heavy to the normal oxygen is $^{18}O/^{16}O$.

Natural Abundance and standards

There needs to be a definition of the 'normal' isotopic ratio for each element, as a basis from which differences can be reported. What is 'normal' is arbitrary, and so is defined by international standards — eg for oxygen 18 the standard ratio $^{18}O/^{16}O$ is defined as the ratio in 'Standard Mean Ocean Water (SMOW)'; for carbon 13, $^{13}C/^{12}C$ is the ratio from the 'PDB *belemnitella americana*'; for deuterium (D or ^{2}H),

D/^{1}H is also from SMOW; for Boron 10, ^{10}B/^{11}B from the 'NBS SRM boric acid standard'

Fraction abundance r

is the isotope concentration of any sample. By convention expressed as parts per mil (‰), not percent. For instance the fractional abundance (r) of carbon 13 in a CO_2 sample is given as:

$$r = 1000 \times [^{13}CO_2]/([^{13}CO_2[+ [^{12}CO_2]) \quad ‰$$

When any process occurs – either reactive or physical – then the new component or phase may differ from its original fractional abundance because of the different rates of change caused by the different masses of the isotopes. Hence the new component can have a different fractional abundance. By looking at the difference in fractional abundance of the different components, we can gain clues as to where these components must have come from, and how they must have formed.

δ Value

The δ value is the isotope concentration relative to the standard, and is the usual way of reporting results from mass spec analysis.

$$\delta^{13}C_{CO_2 (g)} = \{(^{13}R_{CO_2(g)}) - {}^{13}Rstandard)/{}^{13}Rstandard\} \times 1000 \quad ‰$$

where R is the molar ratio of 2 isotopes in a sample

$$^{13}R_{CO_2} = [^{13}CO_2] \ /[{}^{12}CO_2]$$

In words, the δ value is basically the deviation of the isotope concentration from the standard, quoted in parts per mil (‰), and is a particularly easy way of expressing results.

Isotopic fractionation occurs when rates of change depend upon the mass of the element. In simple cases, such as gaseous events, it is possible to calculate the different rates using quantum mechanics, since the energy levels for the reaction rate will be predictable for the larger mass of the element. Similarly, the fractionation due to the vaporization can be predicted theoretically. When a theoretical background is not available, experiments can be made directly in the laboratory to measure what isotopic fractionation does occur.

Once proved theoretically or experimentally, isotope concentrations can be used in evidence of conditions on the planet in its distant past.

Isotope analysis can be used where there have been reactions or physical changes, or equilibria. Equilibria, because equilibria are the result of relative reaction rates, so they are also affected by mass. The elemental isotopes most useful in palaeontology are ^{13}C, ^{18}O, ^{2}H and ^{11}B.

References

Fischer, H., Wahlen,M., Smith, J., Mastroianni,D., Deck, B., *Ice Core Records of Atmospheric CO₂ Around the Last Three Glacial Terminations,* Science , 283,1712, 1999.

Linacre, E., Geerts, B., *Climates and Weather Explained,* Routledge,1997

Monnin, E., Indermuhle, A.,Dallenbach,A., Fluckiger, J., Stauffer,B., Stocker,T., F, Raynaud, D., Barnola, J., *Atmospheric CO₂ Concentrations over the Last Glacial Termination,* SCIENCE, 291, 1125, 2001

Sigman, D. M., and Boyle, E. A.,(2000), *Glacial/interglacial variations in atmospheric carbon dioxide,* Nature, vol 407, 859

Further Reading

Hansen, J., Sato, M., Kharecha, P., Russell, G., Lea, D., and Siddall, M. (2007), *Climate change and trace gases.* Phil.Trans. Royal. Soc. *A,* 365:1925–1954.

Zeebe R.E., and Wolf–Gladrow D., (2001) CO₂ *in seawater:equilibrium, kinetics and isotopes,* Elsevier, Amsterdam

Chapter 9   Sustainable Energy Alternatives

Society has become completely dependent on fossil fuels for the majority of its energy needs. This is bad because this is increasing the CO_2 levels on the planet into uncharted and potentially dangerous levels. It is also bad because fossil fuels will not last for ever, and energy sources are required to last. They must be renewable, or at least able to sustain requirements for a number of centuries. 'Sustainable energy' (and materials) are what is needed to replace the role now taken by fossil fuels. The definition of 'sustainable' is open to debate, but some authors on the subject suggest that if a fuel will be available for 500 years, it can be considered a sustainable solution.

In this chapter we will describe the alternative renewable and sustainable options available. The following chapter will discuss how these possibilities can be introduced into our economy.

We will start by discussing the science of energy, and move on to the science and engineering behind the different energy alternatives.

9.1) What is energy

The study of the scientific principles of energy is a subject called Thermodynamics. This is a most important subject, both in physics and chemistry, and it is taught, repeatedly, as part of scientific and engineering courses. Thermodynamics is not an easy subject and its concepts are difficult to grasp, – energy is an abstract idea, it can take the form of heat, light, sound, motion, kinetic, potential, chemical.......

Energy cannot be created or destroyed but it can be transferred. Matter is a form of energy, which under very special circumstances can be converted to energy forms that we can use. For instance, by splitting uranium, there is a slight change of weight that produces energy as heat. This is the energy provided by nuclear power stations.

Transfer of energy obeys very strict rules. Higher level energies can drop to lower energy levels, but not the reverse. High temperature heat can be transferred to bodies at a lower temperature, but heat will not flow from cold to hot. This direction of flow is quantified by the concept of entropy (S), which is the degree of mixing in a system. Energy might be a difficult abstract concept, but entropy is doubly so. When mixing occurs then the resulting mixture is more random than before mixing. Entropy change must always be positive, which means that whatever change occurs in the universe, the result is the universe being more mixed. All systems move towards being more mixed. If a bag of apples is shaken with a bag of oranges, the resulting equilibrium is a mixture of apples and oranges. However much shaking, there will not be a result when they are again completely separate. Similarly, when a hot body, with its molecules moving fast, is mixed with a cold body with slow moving molecules, the result is a mixture with all the molecules moving around a mean velocity. The system is more mixed.

In heat terms, change in entropy (ΔS) is defined as the heat transferred (ΔH) divided by the temperature (T),

$$\Delta S = \Delta H / T$$

The sum of the entropy changes of all the components in a system must always increase with any change. For any system which appears to go against this law, this is only so because the whole system is not being considered. Heat pumps move heat from cold to hot fluids – apparently contradicting this law, but this only occurs because of the supply of external energy for the pumping. When the total system is considered – the heat change plus the energy supplied – the entropy change is in the correct direction.

So energy change is the dissipation of an ordered system into a more mixed system. There can therefore be no concept of energy recycling – it can only be used up. Once dissipated, the energy is lost to further use. However, energy streams can be carefully routed to cover a series of temperature levels: High–grade energy can provide energy at a high temperature, and the leaving energy stream may well be used for lower temperature duties, and that stream cascaded to even lower temperature duties. This is the next best thing to energy recycling and recovery, and it should always be borne in mind, that the best use of energy is when the appropriate grade of energy is used for the duty required.

In summary; *Energy cannot be recycled, it can only be dissipated. But energy does come in different qualities, and care should be taken not to use high grade energy such as electricity for low grade duties such as heating. When high grade energy has been used, it may become low grade energy, and can used again for heating, rather than being completely discarded.*

In the past, Man's demand for energy was satisfied by the burning of wood and harnessing natural flows – using dams and windmills. Today, the most convenient form of energy is the combustion of the fossilized carbonaceous material which has been deposited by the planet over millions of years. This energy supply is abundant but not unlimited – but it will last for some decades yet.

Problems arise when the amount of CO_2 released into the atmosphere induces changes which may be very detrimental to our future. Whether this is definitely so, or only just a possibility, does not alter the argument that it is reasonable to move away from our reliance on fossil fuel burning as our major source of energy reasonably quickly. Either too much reliance on a single, limited source, or the harmful effects of its byproduct – whatever reason – we need to investigate other sources of energy. We must be searching for alternatives.

There is abundant energy for our needs radiating down upon the planet from the sun; it is a matter of harnessing this energy, keeping our energy demand in balance with the energy supply. To harness this energy we must capture the incoming radiation either directly, or let it be captured by biomass, or capture the motion induced by the unevenness of this radiation – from wind, sea movements and river flows. These energies are truly renewable, but they have the drawback that they are all intermittent and cannot be centralized into a few large, economic installations.

The only other source of sustainable energy available to us comes from converting mass into energy – this means using some form of nuclear fission or fusion. This is already supplying 15% of our energy needs, but there are serious questions as to its safety and security, and it is generally out of favor with the public.

Now let us look at the various alternative energy sources that are available in three different applications – power, transport and heating.

9.2) Alternatives for Power Generation

The generation of electrical power is normally done by driving electrical generators, which requires a supply of kinetic energy to spin these generators.

This kinetic energy required comes traditionally from steam power, the steam being raised mainly from fossil fuel burning. To use the same system, but to burn renewable biomass material such as wood or vegetable matter or use nuclear fuels means we have achieved the aim of generating power without the use of fossil fuels.

The perceived wisdom for power generation is to concentrate production to a few large plants. This means, for instance, that biomass must be collected over very wide areas and the cost of its transport to a centralized location becomes one of its disadvantages. In fact most generation from renewables has the same disadvantage that large centralized plants are no longer possible. To try to keep the concept of large scale electrical production without CO_2 release to the atmosphere, one solution is to keep with fossil fuels but to capture the CO_2 produce and store it permanently away from the atmosphere. This is called 'Carbon Capture and Storage' (CCS) and it is an agreed policy that is expected to apply to any new fossil fuel plant built in the future.

The generation of electricity by wind requires thousands of small generation plants with an average production of 0.5 MW, linked into the grid system. There would need to be 16.000 generators to replace one centralized plant. This may be viewed in horror by traditional power generation engineers, but we must come to accept that 'distributed' power generation systems, feeding into an established electricity distribution grid network, may be the way forward if we are going to use more renewable energy for power generation.

There is also the possibility of photovoltaic cells (PV) for generation of power directly from the sun's radiation, where light is converted into low voltage electricity by voltaic cells and electronic equipment is now available to lift the low voltage generated to a voltage suitable for feeding into the grid. Again this is very much a 'distributed' generation system.

If all else fails, there is always nuclear energy to fall back on. Its uranium raw material is abundant; it has large centralized plant and is therefore economic; it can work with existing infrastructure; it creates no CO_2. The disadvantage being the unsolved problem of the long term radioactive waste for which there is no agreed solution and and security risks, as it involves the raw material used in nuclear weapons. Additionally, there is a strong public fear of nuclear energy, which makes policy–making very difficult in a democratic society.

Now let us look in more detail at these possibilities that are open to us.

9.2a) Hydro Power Plant

Hydroelectric power is the most obvious renewable energy source for generating electricity. Unfortunately most opportunities have been exploited, and there is very little chance for new hydroelectric power plant to be built in the developed world, as almost all suitable sites have been developed. There is opportunity for the poorer counties such as Africa to develop hydro power, but they are hamstrung by its high capital cost.

There is, however, still opportunity to harness much smaller amounts of energy in rivers and flowing water. This is designated ROR – Run on River, and consists of installing turbines and weirs in running rivers to extract some energy from the flow, returning to the old idea of water mills. The generation can be economic, but it is only small scale. People living in old mills find this an attractive proposition, but the numbers are so insignificant that it cannot contribute seriously for national requirements, but for individuals it can be very worthwhile.

9.2b) Biomass

Bioenergy is a term covering a wide range of processes in which energy can be recovered from biological matter, usually crops. Since crops grow annually, it is indisputably renewable energy. Simple burning of wood or agricultural or domestic wastes can be used to raise steam to generate power. This is a very suitable way of raising energy if the biomass is readily available and has to be disposed of somehow. As a major contributor it fails because of the quantities involved, the difficulty in collection and transport, and the damage it can do in over–harvesting. At the local level biomass may sometimes be financially attractive, but it is no solution for major energy supplies.

The burning of municipal waste is sometimes included under biomass. Here the material must be disposed of. It can be burned to generate steam and hence some power, and the waste heat from the turbine can then be used to provide district heating (combined heat and power (CHP)). An alternative municipal waste process subjects the waste to very high temperatures (pyrolysis) to break down the organic material into combustible gaseous products which are then burned to generate power.
Such proposals can be economic, as they perform a multitude of tasks, but their contribution to power generation will not be significant.

Biomass is a more significant fuel for transport and heating.

9.2c) Wind power

Using wind turbines to create electric power is now well established, supported by government policy and incentives for their installation. Large turbines rated at 2 Mw are now being built. They are normally arranged in wind farms which are carefully located where they will receive most wind. Figure 9.1 shows a general view of a wind farm. The major problem is the large number of these wind generators that must be installed to create a noticeable contribution to our energy supply. The second problem is that they are rarely working at name–plate capacity. They produce about 25% of their installed capacity averaged over the year. This matter of intermittency will be discussed in detail in the next chapter.

Figure 9.1   Wind turbine farm
With the average output of each wind turbine being only 1.0–2.0 MW, many thousand must be erected over large tracts of land to create a meaningful contribution to the power generation requirements. With each turbine requiring planning permission, and little public support for them, it will be difficult to erect the required number on land.

Since suitable sites for onshore wind farms may become difficult to find, as they are often objected to by residents, there is some attraction in building wind farms in the sea – offshore. These wind farms can be more extensive; there can be more machines, the machines can be bigger and the achieved availability is about 30%, which is 5% better than onshore. Unfortunately, there are extra costs involved in construction in the sea and maintenance at sea, making offshore considerably more expensive than onshore installations.

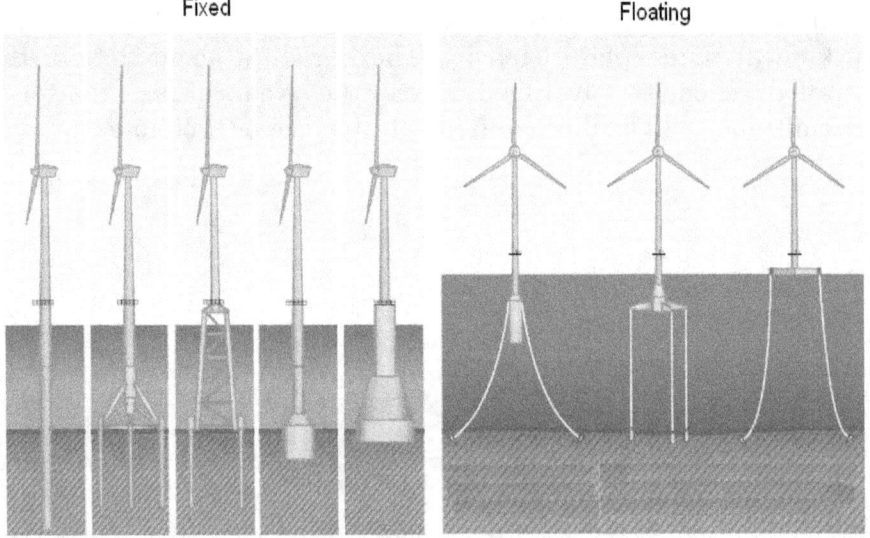

Figure 9.2 Off–shore wind turbine design *Off–shore wind turbines overcome the problem of siting and planning permission, and enable larger turbines to be constructed. The problems of supporting them on the sea bed and being robust enough to withstand violent storms and corrosion, causes design problems and increased costs which make them an expensive alternative to land–based turbines.*

The size of the wind generators has steadily increased, with sizes of 2MW for onshore, and 3MW offshore being attainable at the present time. Figure 9.3 shows diagrammatically, the history of the growth of these turbines, and gives an impression of the future possibilities.

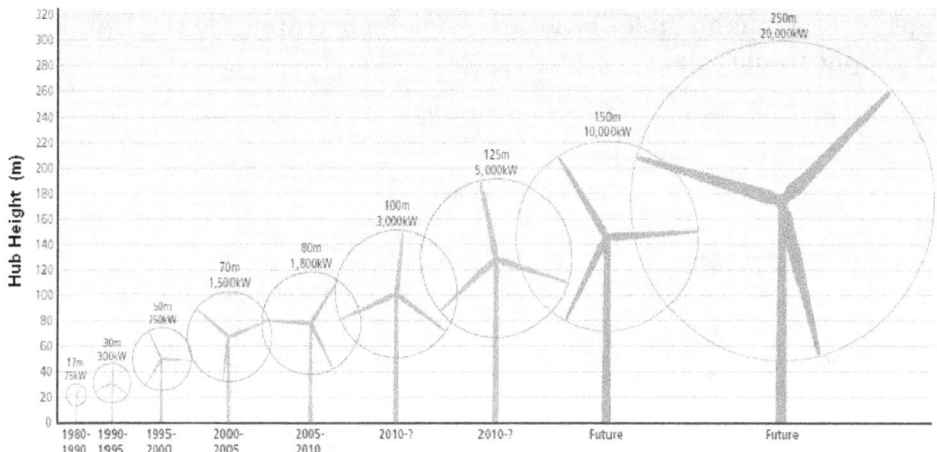

Figure 9.3 ***The growth of wind turbine size*** *Individual wind turbines produce only small amounts of power. There is a need for each unit to provide more power to contribute meaningfully to energy needs. With time the unit size has increased, and future sizes are expected to be even larger. Public opinion is against large turbines, and so the largest will be employed off–shore.*

9.2d) Solar power

Solar *thermal* energy is the capture of sunlight directly as heat. This can be done with simple panels, with water flowing over a blackened plate, and back into warm water storage. This is a very simple form, and has been used for centuries in sunny climates, a very popular form involving a board and an oil drum.

The method can be refined, so that the heated fluid can be passed through pipes which are vacuum– jacketed, so there is no heat loss. Temperatures of $50 - 120\ °C$ are then possible. Further developments involve the concentration of the sun's rays onto a central heat exchanger which contains a heat transfer fluid. This fluid can reach very high temperatures and by using a molten salt, $400\ °\ C$ can be achieved, and this used to raise steam and generate power. With such schemes the mirrors concentrating the sunlight are computer controlled in 2 axes, so that they follow the sun and maximize the capture of sunlight. Installations have demonstrated the concept, but few applications exist.

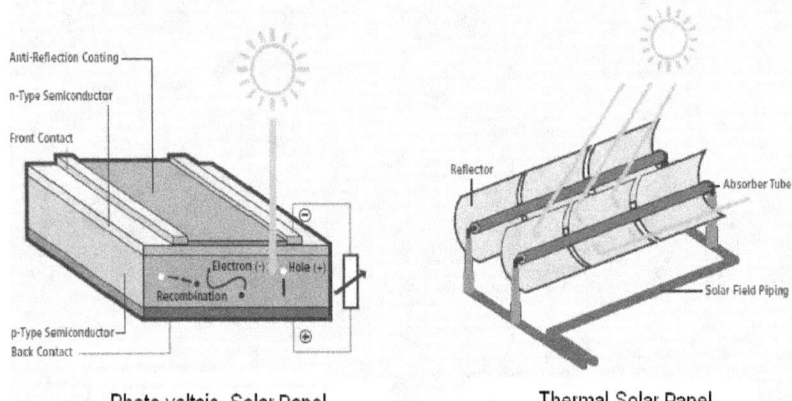

Photo-voltaic Solar Panel   Thermal Solar Panel

Figure 9.4 *Photovoltaic and Thermal Solar Panel construction principles.*

The two types of cell are fundamentally different: Photovoltaic cells are semiconductor devices which generate electricity and after voltage increase can be used in the national grid: Thermal solar panels simply warm water by absorbing the sunlight, and are usually used to provide warm water for domestic use.

Solar *photovoltaic* cells have become the established way of capturing solar energy, and converting it directly into electric power, which can be fed into the national grid system. These systems are well developed and, again through government incentives, are now widespread. They are reputed to be trouble–free, long–life, with minimum maintenance; they simply stay there feeding the electricity grid with energy for years. They are small scale, each panel being about nameplate 1.0 kW. They operate at nameplate capacity only when the sun is directly overhead with a clear sky. This means the annual availability in reality is only about 10% of the nameplate capacity in temporate climates, so large numbers must be installed to contribute significantly to the national demand. Being small scale they are expensive to install per kWh produced. They are very suitable for supplying small power requirements at remote locations where no cabled power supply is available.

Figure 9.5 Large Thermal Solar installation *Situated in the desert, these two installations of tracking mirrors concentrate the sun's rays on a target at the top of the towers. The target reaches sufficient temperature to raise steam to operate the steam turbine generators which are part of the plant at the base of each tower.*

Figure 9.6 Large Photovoltaic Solar Installation
Again situated in the desert, this installation of photovoltaic cells demonstrates that, given the right conditions, such installations can provide a credible source of electric power.

9.2e) Wave Power

It has been estimated that there is sufficient energy available in the oceans to satisfy all our energy needs, There are 6 distinct ways of extracting energy from the oceans. but only one of the methods has yet been developed to the stage of commercial application.

The six possibilities are:

Wave energy   – where the wave motion brought about by energy from wind is captured by small devices which nod or rock with the waves, and this movement is converted into motion to drive a turbine.

Tidal range   – where energy from gravitational forces causing the tides, is captured.

Tidal currents – where energy from tidal flows in estuaries is captured by damming or 'barraging' the estuary at high tide, and letting the water return to the sea at low tide through turbines.

Ocean currents – where huge turbines are submerged in stable ocean currents, and the water passing through them generates power.

Ocean thermal energy – where the difference in temperature between the cold deep and the warm surface operates a reverse heat pump/turbine with an intermediate fluid such as ammonia. An idea borne through thermodynamics. The energy is there, but it is difficult to visualize how the plant could be constructed in the deep, tropic ocean.

Saline gradients – where the osmotic pressure difference between the sea water and the fresh river water entering the sea can be used to drive a turbine. Again a theoretical idea to capture energy which is there, but difficult to visualize as operating plant.

Apart from an example of the estuary barrage, there are no commercial plants using wave power. Numerous different devices have been proposed, but their installation by the thousands to capture reasonable amounts of energy has never been realized.

It is claimed that wave power is more reliable than wind because the tide is there twice every day, whereas wind can disappear for weeks at a time. This is true, but energy production twice a day, and even then not a constant amount, does involve solving intermittency problems similar to wind power generation.

9.2f) CCS – Carbon Capture and Storage

The concept of carbon capture and storage (CCS) is to use existing fossil fuels, but to capture the CO_2 in the fuel gas and to store it indefinitely. This suggestion is taken very seriously by planners because there is difficulty in matching future energy demands if no more fossil fuels are to be used. The IPCC has produced a 400 page report on the possibilities and so we should look carefully at what is involved (IPCC, 2005)

The process of capturing CO_2 from flue gas at a 20% CO_2 concentration in N_2 is technically achievable using an amine absorption and stripping process and has been operated for many years. An amine absorbs the CO_2 in one tower and this is then distilled off in a second tower to release the CO_2 for compression and storage. The amine process is used to remove CO_2 from H_2 and a similar process is used to remove H_2S from refined fuels. The process is extremely energy intensive, and an alternative process is to use pure O_2 in place of air for the fuel combustion, the flue gas is then only CO_2. A third proposal is to use the fossil fuel to produce H_2 and CO_2 by steam reforming, and then remove the CO_2 from the H_2 by the amine process, which again is known technology.

Once the CO_2 has been isolated, it can be compressed or liquified and sent for permanent storage. The most likely form of permanent storage is to inject it into spent oil or gas fields or into saline formations where an impervious upper layer prevents the CO_2 escaping into the atmosphere. Geologists are fairly certain that the majority of the CO_2 will remain captured with losses of about 5% in 100 years, as long as the facilities are 'maintained' ie when escapes are noticed, the escape route is plugged. There have been a small number of trial storages, demonstrating the principle and now (2012) a number of further test cases are being proposed – particularly in Australia. Other permanent storages suggested are that the CO_2 could be returned to the bottom of the ocean, from where, over 1000 years it would gradually mix with all the sea water and would produce only small changes in sea water composition. Another suggestion is that the CO_2 be mixed with crushed silicate rocks where the weathering reaction will produce $CaCO_3$, $MgCO_3$ and SiO_2, absorbing the CO_2 – called mineral carbonization.

Figures 9.7 and 9.8 shows diagrammatically the possible ways of CO_2 disposal.

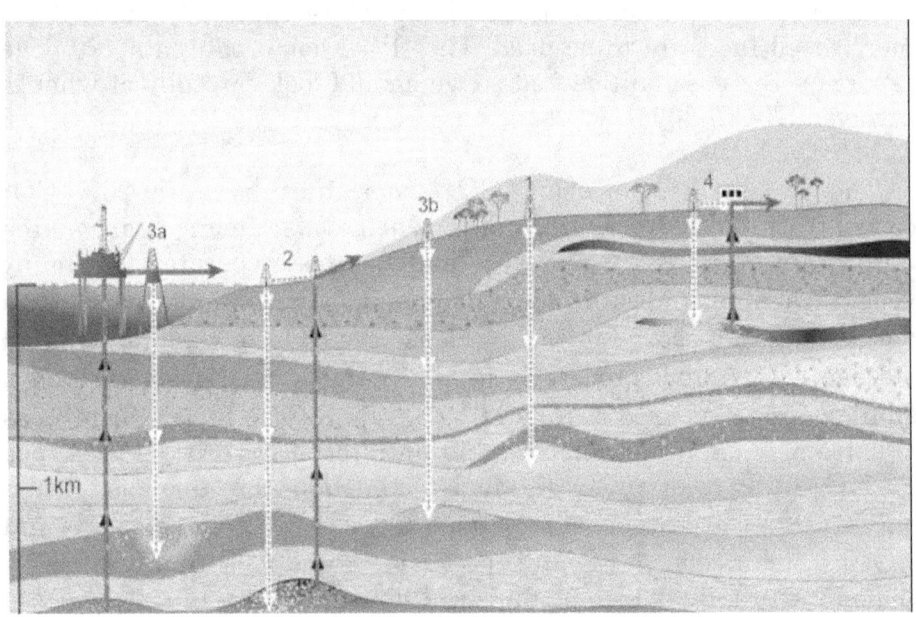

Figure 9.7 CCS Geological storage Possibilities
Credible long term storage of high pressure CO_2 in geological formations associated with expended oil and gas wells, or deep saline formations is possible. CO_2 can also be used to enhance the recovery of oil and gas, from partly exhausted wells.

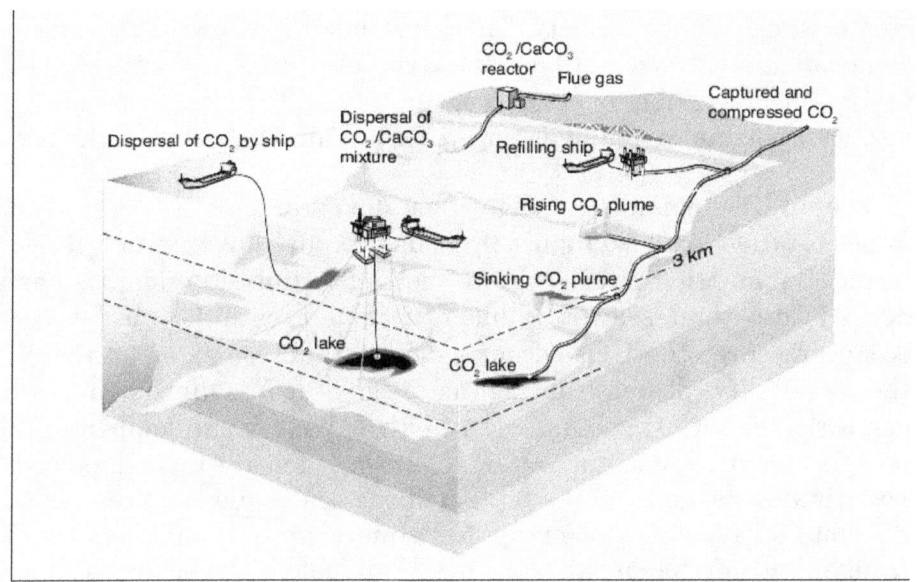

Figure 9.8 CCS undersea Storage Possibilities
Compressed CO2 is less dense than seawater at pressures equivalent to 3km depth. Below this depth the CO2 is heavier than seawater. This leads to the two suggestions; that collected CO2 could be distributed above 3 km to be absorbed in the sea; or pumped to greater depths, where it should remain as a pool of CO2, mixing very slowly with the bulk of the ocean. None of these suggestions has been tested.

The IPCC report goes into great detail of methods of transportation, costs of the process, transport and storage, to show that these possibilities are feasible.

CCS uncertainties

CCS has certainly a number of disadvantages. Firstly, 10% more gas and up to 40% more coal is required to provide the energy for the separation and disposal of the CO_2. This is a wanton consumption of fuel and organic raw material which will be at some time in short supply.

Secondly, the storage underground is untested and will be no more attractive than the burial of nuclear radioactive waste, which creates much public anxiety. Though it may be shown by extensive demonstration projects that this form of storage is possible, the geographical location of the power plants and the CO_2 reservoir will also be problematic, requiring pipelines or shipping between the two.

The other remaining suggestions for storage – oceans and mineralization are also highly problematic, and have never yet been tested. Submerging in the oceans seems fanciful. Liquid CO_2 has a density less than seawater under normal pressures, but at deep ocean pressures the CO_2 becomes denser than seawater, so there is the possibility of distributing as a gas at medium depths, or deposition as a CO_2 pool at great depths. What would happen to the pool, adjacent to reactant water? Will some density inversion occur? Distribution as a gas on the bottom of the sea would require monster distribution networks of many 100 square km. It all seems to be fraught with problems, all to be overcome before the schemes can be successfully developed. The mineral carbonation requiring the mining, crushing and solution of million of tons of rock, many times more than the fossil fuel used, using a notoriously slow reaction, is as equally unlikely to come to realization as using the oceans to disperse the CO_2.

Nevertheless CCS is being considered very seriously. At least 10 CCS storage trials are being made and most long–term energy plans usually feature CCS as a major contributor to our future energy mix.

9.2g) Nuclear Power

The Science

There are 92 stable elements on the planet. More elements are known, but these are unstable and break into smaller atoms. Even the larger stable elements are breaking down – they are radioactive elements such as uranium and radium – and as they break down, mass changes occur, and energy is released.

There are stable isotopes which absorb neutrons and produce unstable isotopes which lead to further reaction – these isotopes are called fertile, because these neutron absorptions set off a chain of nuclear activity.

Some isotopes are very unstable and emit beta rays (electrons), so converting to other elements.

Some isotopes are so unstable they break up into two smaller elements and emit more neutrons with the liberation of much energy as their total mass changes. These are described as fissile.

Nuclear energy is obtained by finding a sequence of atomic reactions which start with fertile material and lead to fissile material with the consequent liberation of energy and sufficient neutrons to maintain the reaction chain going until the fertile fuel is spent.

There may be many possible combinations of elements that create such a chain, but two reaction chains are well known and have been used in commercial–sized reactors; Uranium and Thorium.

Table 9.1 shows the relationship between the elements and their isotopes at the heavy end of the Periodic Table. Table 9.1 also shows the routes taken by the uranium–based and thorium–based reaction chains from the fertile to the fissile material which produces the energy.

Uranium is interesting because one of its naturally occurring isotopes ^{235}U is radioactive and slowly decays, emitting neutrons. These neutrons collide with the fertile uranium isotope ^{238}U to produce plutonium which is fissile, which splits into smaller elements emitting the energy and more neutrons. These neutrons collide with more ^{238}U, the chain continues and energy is continually released. Each neutron collision cycle must produce more than one neutron – to maintain the reaction chain and to compensate for neutrons absorbed by other elements not in the reaction chain. If fission produces too many neutrons, then each neutron produces more fission and the result is an explosive release of energy. If the number of neutrons is controlled or moderated, then the heat release is steady, and can be used for power generation.

Table 9.1   Tranformations between heavy elements isotopes
showing the Uranium and Thorium reaction chains
($\rightarrow$ denotes beta decay by ejection of an electron)

Atomic wt	Th 90	Pa 91	U 92	Np 93	Pu 94	Am 95
241						
240						
239			↑→ →	→ →	☼fissile	
238			♀fertile			
237						
236						
235			☼ fissile			
234	↑ → →	→ →	☼ fissile			
233	♀fertile					
231						

But uranium is not the only possibility. There are other nuclear reactions possible with different elements which can liberate heat on collision with neutrons, and sustain the reaction by breeding more neutrons. One such element is thorium as shown in Table 9.1.

The energy arises through the splitting of the fissile element, losing mass and therefore liberating heat in the process. This heat is then captured by a heat transfer medium such as water or gas and used to generate steam and finally electricity. About half the energy becomes electricity and the remaining half must be dissipated in cooling towers or the sea – always the unfortunate consequence of generating electricity from steam.

Uranium based Nuclear Power

Uranium fueled nuclear power is an established technology which produces electricity on the large scale and does not involve the production of CO2. Nuclear power is more widespread than is generally appreciated. Probably three quarters of the world are using it, or planning to do so; Africa, Australia and the Middle East being the only non–nuclear powered areas at present. France produces 75% of its electricity by nuclear power, and about 15% of the world production of electricity is by nuclear fission.

There have been different stages of atomic power reactor design. Generation 1 is now obsolete, being replaced by Generation II (LWR and BWR for example) which are now in operation.

Generation III are the newer type, about to be built with longer lifetimes eg 120 years. Generation IV are being developed and are expected to be in operation in 2020. The early reactors employed collision of slow neutrons. The later generations are fast breeders using fast neutron collisions, which require less enrichment of fuel, consume much more of the uranium, and are capable of fission with a much wider range of fuels.

Generation II and III enable reprocessed uranium and plutonium to be recycled in the reactor fuel. Generation IV will enable other reprocessed radioactive elements from the nuclear waste to be recycled.

Thorium – an Alternative Fuel for Nuclear Fission

The accepted fuel for nuclear reactors is uranium, because this was developed in conjunction with the military need for weapons. The enrichment of uranium, and the production of plutonium by the reactors fitted the military need as well as producing the design of the first commercial reactors. All development has gone down the uranium route, and to consider alternative fuels is not economically attractive because of the need to return to development work, in an industry wanting to proceed with a process it has already developed.

Thorium is a heavy element, which is capable of nuclear fission, and experimental thorium reactors have been built. Thorium will not sustain reaction without being seeded with uranium, but it is considered to have as much potential as the developed uranium route to power generation. Its costs have been estimated as being roughly equivalent to or cheaper than uranium fed power generation; thorium has been estimated as being 4 times as available as uranium; there is a reduced radioactive waste problem, and it does not produce plutonium, so reducing terrorist security risks. Above all, it is an alternative nuclear fuel, so reducing the problems of security of supply, which could be a major concern if uranium fuel were the only nuclear fuel available, and nuclear power developed to be a major player in the future.

Supporters of thorium technology will claim the only drawback with thorium is that the necessary detailed development work has not been completed. Development for uranium was carried out because it was financed by the military need for plutonium. Development costs for a thorium process are too much for private companies to contemplate, but national sponsored development programs are underway in China, India and Canada.

Nuclear Waste Treatment

Bombarding material with neutrons is not an exact science, and many unstable elements are produced which are radioactive. Radioactive material is the result of the fission producing the energy; there will also be fissile material such as unspent ^{235}U, and ^{239}Pu, which remain radioactive for thousands of years, together with many other elements which have been produced by the random bombardment.

After some years in the reactor the fuel rods are expended and have to be replaced. The old rods contain a mixture of elements, some of which are highly radioactive. These rods are the radioactive waste from the process. At present there have been 300,000 tons produced in the world and future amounts are expected to reach 1,000,000 tons. How to handle this waste is a main drawback to nuclear power acceptance.

The fuel is 'spent' when the rods contain sufficient reaction products which absorb neutrons to prevent the chain reaction continuing. This spent fuel contains about 94% ^{238}U, 1% ^{235}U, 1% plutonium, with 4%

fission products such as actinides and lanthanides. The radioactive components uranium and plutonium can be recovered in Reprocessing Plants and recycled by incorporation in fuel rods as mixed oxide fuels (MOX). In this way the radioactive wastes can be reduced to 10% of their initial volume. Though it is sensible from the point of sustainability to recover the uranium and plutonium for recycling, the processes are at present very expensive, more expensive than burying all the spent fuel and just using new, freshly mined uranium. This means that some first attempts at MOX processing have fail on economic grounds. There needs to be more development work on the processes, a higher price for uranium, and a greater demand from newer generation reactors before waste reprocessing becomes well accepted.

Newer waste separation processes are being developed to isolate and recycle other radioactive products, the actinides, transitional elements with atomic weights from 89 – 102. Of these, Curium, Cm, and Americium, Am are radioactive elements which can again be incorporated in new fuel for Generation IV reactors. The problem lies in a clean separation of those materials which can be useful as fuels, from those elements which will simply absorb neutrons and quench the reaction. The actinides are capable of being incorporated in fuels, but the lanthanides, transitional elements with atomic weights 57 – 75, are net absorbers of neutrons and so would quench any nuclear reaction.

The processes required in these Reprocessing Plants are complex liquid /liquid extractions, choosing a carrier which selectively separates just the required elements from the waste fuel. The spent fuel rods are dissolved in nitric acid and the required elements are extracted from this aqueous phase into a kerosene phase by very specific extracting agents. Extracting uranium and plutonium uses tributyl phosphate in the PUREX process. But reprocessing involving the separation of the actinides Cm and Am require more complex ligands, which is an active area of research (Taylor, 2011). When the fourth generation of reactors are in operation, together with using these new recycled materials, the resulting waste will be only a few percent of the radioactive waste from the generation I reactors.

Reprocessing vs Long–term Storage

At present, only 30% of the radioactive wastes have been reprocessed. Reprocessing is expensive, and it is cheaper to accept medium term storage as a solution for the waste. Medium term storage for decades under water is a necessary first step to remove the heat from the short– lived fissile material which generates too much heat to consider immediate long term storage. The long term storage, for thousands of years, is being actively discussed where the present thoughts are that of calcining it in a glass and deposition in deep burial sites.

Opinions differ on reprocessing because reprocessing plants recover fissile uranium and plutonium, which represent serious security risks should they get into the wrong hands. In nuclear waste these elements present no risk as their handing and separation are so difficult. The US for instance no longer does any reprocessing, and would rather that there was no reprocessing being done at all, so giving greater security.

It is not so much that there is no solution to the nuclear waste problem, as that the solution by recycling is being hampered by security policy and economics.

Nuclear Fusion

Nuclear fusion is the combination of atoms which collide with such energy that new larger atoms form from particles collisions – for instance deuterium with deuterium, or tritium with deuterium to form helium. There is a mass change in the process and this is liberated as heat. This is the process occurring on the sun, and is the principle behind the hydrogen bomb.

Hydrogen is the usual element used in the collisions, and the energy released is huge. So this promises an unlimited energy source. The technical problems for its commercial realization are the containment of the intensely hot fusing media, and the extraction of heat from it. Temperatures in millions of degrees mean that no material can be used to contain it – it has to be suspended in a magnetic field, away from any walls. Then heat must be extracted. This process has been discussed and investigated for over 50 years, but we seem to be no nearer harnessing this energy on a commercial scale. It cannot be considered in future energy planning because there is no knowing if the immense technical problems will ever be solved.

9.2g) Geothermal Generation

Geothermal generation is the generation of power from the heat from the earth's interior – see figure 9.9. Sufficiently high temperatures are normally found at depths around 3 – 10 kilometers below the earth's surface. In geophysically unstable zones, this energy is closer to the surface and shows itself as geysers and general volcanic activity. Hence the energy is more readily available in areas such as the Pacific Ring, where the tectonic plates meet.

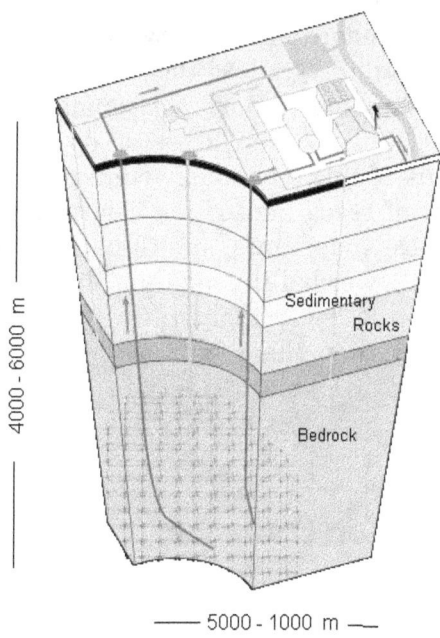

Figure 9.9 Geothermal Power Plant
Water is pumped in deep borings to reach temperatures of around 300°C. The resulting steam returns to the surface where it is cleaned and fed to power generation turbines. The plant is economic if the heat is near the surface, as is the case in volcanic areas or where hot springs are found.

This energy source is rather different from the much shallower boring to obtain a heat source for domestic heat pumps. In this case the object is to obtain an adequate heat source at constant temperature regardless of season, for lifting up to domestic heating temperatures. In geothermal energy temperatures of 300° C are normally required to create steam for driving steam turbine generators; heat pumps function with 10deg C. Some purists do argue that this is also geothermal heat, as the energy does come from the same source. This is true, but the processes are so different it is best to discuss them separately.

Geothermal energy is a very practical energy source of supply for areas with tectonic activity. For instance between 10 – 20 % of the power requirement of Iceland, the Philippines, and New Guinea are generated in this way. The Philippines and US generate over 10TWh/y each – a considerable quantity. The boring costs make this an expensive form of energy, unless the heat is near the surface, and it is not considered economic in normal circumstances.

The complications of this technology stem from the cleaning of the steam and removal of trace impurities before it is fit to be fed to a steam turbine. Developments of the technology include the possibility of using a lower boiling medium than water – eg butane or propane – to operate at lower temperatures than those required by water. This would reduce the depth of boring required to attain workable temperatures (Dobbie, 2011).

9.3) Alternatives for Transport

Apart from some trams, trolley buses and a few trains, the fuels used for transport are entirely fossil based. There is good reason for this. These liquids fuels are very easy to handle and transfer. They are very compact and enormous quantities of energy can be transferred in a few minutes by transferring the liquids. Now they are so universally used that an entire supply chain is complete over the whole world, and of course, all vehicles have been designed for their use, and so vast investment has been made in them.

Let us look at the various alternatives to fossil fuels that are available for transport.

9.3a) Biofuels

In finding a replacement transport fuel, the first option is to look at replacing the fossil fuel with a non–fossil–fuel–derived liquid. Hence the attraction of biofuels which are liquids derived from plants, and plants are renewable. Of these, ethanol is the longest established and this has been used for transport fuels for the last 60 years. Most well–known is the Brazilian policy of fermenting sugar cane to provide ethanol to reduce the country's dependence on imported petroleum, and the concurrent saving in foreign currency. There are further instances of ethanol being produced from wood or waste liquors from paper pulping plants which were used in Central Europe during World War II. There is much waste cellulose material – wood and other ligno–cellulose materials – and since cellulose is a sugar, it could be expected to be a suitable raw material for fermentation. The difficulty is in hydrolyzing the long chain cellulose sugar to simple sugars economically. Research is well advanced in developing enzymes which can breakdown the cellulose, leading the way to a large amount of waste biomass becoming suitable for conversion to ethanol–based fuel.

An interesting ethanol project has been announced where over 600 small ethanol plants will be built in Indonesia and Malaysia to convert the waste fronds from date palms into ethanol to be used in transport fuel. Interesting, both because it is a courageous use of technology on the small scale, and because the project is being funded by the Chinese.

The second important form of biofuel is vegetable oils extracted from plants. Plants have always been grown for oil extraction, and this oil is comparable to diesel in its properties, and so can be used in transport. A variant of this is the collection and use of old cooking oil as diesel fuel. Though very newsworthy when done in small local workshops, it is hardly a solution to the future transport fuel problem because the quantities are so small.

Bio fuels can be derived using the organic nature of biomass. 'Pyrolysis' is the name for heating the complex organic molecules of the bio–material to break them down into a range of simpler organic chemicals, which can be converted by various chemical processes into liquid fuels for transport. Very extreme pyrolysis conditions can be used to get very simple molecules. For instance a Westinghouse plasma gasification process, using steam, oxygen and heating to 5000 deg C, is a process for treatment of domestic waste. The products are hydrogen and carbon monoxide, and a glassy slag which can be used as building material. The gases can be burned to generate electricity, or converted into liquid transport fuel.

The major problem with biofuels is that they are in competition for land use with food production. They can be useful as a local tool to enable unused land to be productive, but on the global scale it is fairly clear that with population increasing there will always be a pressure on the land to produce food. The idea of rich nations using their excess land for transport fuel production, whilst other nations starve is not a socially acceptable prospect, and so it is unlikely that biofuels will become the replacement for fossil fuels in transportation.

Biofuels are however finding a useful application in producing an immediate reduction in CO_2 release, simply by mixing 2 –3% biofuels with fossil fuels. Politically speaking this enables target reductions in CO_2 release to be achieved, but technically speaking it is rather dishonest, by not really addressing the real problems of the future.

9.3b) Hydrogen

The replacement transport fuel most seriously considered is hydrogen. This is attractive because it is clean, in that the product of its combustion is water, which is a great relief for cities where transport pollution is always of concern. Secondly it is a great energy carrier – providing a great deal of energy per unit mass; Thirdly it is a fluid; it can be stored and transported relatively conveniently, although there are serious safety problems to be addressed.

Although it can be used directly in an internal combustion engine, simply replacing the hydrocarbon vapor, with very little adaptation of the engine, this is not likely to be the way in which hydrogen will be used. Its real attraction is because it can be used in conjunction with fuel cells to generate electricity to power the vehicle. This combination of fuel cell and electric motor has a higher efficiency than using the hydrogen in an internal combustion engine, and is simpler to manufacture. A fuel cell is similar to an electrolytic cell, but works in reverse. By feeding hydrogen to one compartment and oxygen to a second, divided by a membrane, these two gases produce water plus an electric current between the electrodes. Hence there is the potential to feed hydrogen and air to the cell and produce electricity to power. the vehicle.

A major concern with hydrogen as a fuel for transportation is its handling. It can be stored under very high pressures such as 700 bar, in cylinders and these mounted on the vehicle. It can be stored on a medium such as a metal hydride which absorbs large quantities of hydrogen; this also being contained in cylinders. This absorbent effectively reduces the pressure needed for adequate storage. Other storage methods have been considered, eg methylcyclohexane, which can be dehydrogenated to toluene on board the vehicle, but this requires too much organic liquid to be carried around to provide the necessary quantities of hydrogen to be practical.

The hydrogen has to be produced and at present hydrogen is produced by steam reforming using fossil fuels. To produce hydrogen without fossil fuels, can be done electrolytically, rather than chemically as it is at present, by electrolyzing a weak solution of alkali such as sodium or potassium hydroxide, to produce hydrogen and oxygen, by feeding water to the cell. There needs to be development of the cells and their membranes for large scale production of hydrogen, as the current density of present cells is very low, requiring large volumes of cells to provide useful quantities of hydrogen.

An attractive concept is the local production of hydrogen at each filling station. The station would need no deliveries as the power required for the electrolysis comes from the electricity grid. There would be no need for transporting or even holding large quantities of hydrogen. It begins to look even more attractive than the present fossil fuel distribution system.

A hydrogen economy achieves the aim of moving away from fossil fuels when the electricity for the electrolysis is produced from non–fossil fuels. This does mean more nuclear power and wind farms would be needed to cover this extra power requirement.

Fuel cells have been well developed because they have been used in the US NASA space programs as the means of power generation in space, so the research has been done. That most car manufacturers have made operating prototypes indicates that the technology is well advanced. There have been over 30 experimental vehicles produced.

9.3c) Electric Vehicles

Finally we have the possibility of powering all transport by electricity directly, and this is the most likely direction for the near future. Electrically–powered vehicles were used in the 1920s to 1940s as local delivery vans. For local delivery vans it is ideal. Distances traveled are small, and never far from the charging point. For it to be used as a general fuel for means of transport it must be able to go much longer distances at much higher speeds than our delivery van. This requires more electricity to be stored on the vehicle between charging stops. The problem is storing this charge, which used to be done by lead accumulators. The kWh capacity of a lead accumulator is too low for general transportation, leading to the battery needed being too heavy. The widespread use of electric transportation depends entirely on developing batteries with a higher capacity.

The lithium–ion battery does seem to be the most suitable at the present time, and a number of car manufacturers have lithium–ion based electric cars on the market (2012). Since their major drawback is still the quantity of charge they can hold – and hence their range between charging points –   these vehicles at present are most suitable for shorter journeys, such as commuting and city driving. Other drawbacks are the long charging times required eg 5 hours for a full charge and 30 minutes for an 80% recharge. Battery life may also be a problem. The battery can fairly quickly lose 10% of initial capacity – reducing further its range.

However, electric vehicles are here to stay, and battery developments will very likely reduce these current problems considerably.

As with the hydrogen–powered vehicle, for it to be fossil fuel independent, its electricity must have been generated by renewables or nuclear power.

Hybrid cars are now available and very popular on the market. These are not attempts to move away from fossil fuels but they do reduce CO_2 emissions as they are more efficient. Reducing speed by braking is simply changing the expensively created kinetic energy of the vehicle into waste heat in the brakes. If this braking energy is recovered, then the consumption of the vehicle is better. If braking is done electromagnetically and the braking generates electricity, then this energy can be recovered. The wheels are fitted with generators which are used to reduce the speed.. This requires the car to have a secondary electrical power system to reuse this energy. Hence the name 'Hybrid'. It is powered by conventional internal combustion engine and secondary electric motors. To achieve adequate storage of the braking energy the battery size has to be increased. This arrangement also enables the inefficient operating range of the internal combustion engine to be overcome:– for instance slow movements in towns can be switched onto the stored electric power, which has been generated by the engine operating at its optimum efficiency.

This system produces improvements in consumption of the order of 30% and so is a very attractive marketing proposition. It is, by itself, not an attempt at moving away from fossil fuels.

The concept of plug–in hybrids does go some way to reducing the fossil fuel consumption – by having a larger battery and the ability to charge this from an external power supply, as long as the power has been generated from a non–fossil fuel source. By having an on–board internal combustion engine, and a liquid fuel, when the battery is exhausted, the engine can be used instead of having long charging times at inconvenient times, or worse, being stranded away from a charging point, so overcoming the major drawbacks of a purely electric vehicle.

Another possibility is to have battery swapping at service points – rather like the stage coach changing horses every 15 miles. This has been proposed for fleets of municipal buses, but to have a nation wide service would require a system of physical lifting gear, considerable standardization of automotive design, and a complex financial accounting system.

9.4) Alternatives for Heating

Domestic heating, as opposed to power generation and transport, can be much influenced by making savings. A second characteristic is that smaller, more local solutions are possible – particularly regarding biomass, which has been the traditional means of heating, and still today is the predominant domestic fuel when considered worldwide.

9.4a) Reducing heat loss

A considerable proportion of the energy required by mankind, particularly in developed countries is that required for domestic heating. From a technical point of view this is a very undeserving cause. The need is to keep ourselves warm. This can be done by wearing more clothes, as is done in simpler societies, rather than by warming our living space to about 20–24°C. If we choose to warm our living space, once warm, it   requires no more heating except to make good the heat escaping to the outside world because of incomplete insulation. Policies for reducing energy demand for domestic heating therefore concentrate on improving insulation.

The energy required for domestic heating is low grade heat (low temperature), which is generally considered as waste heat in industrial situations and which poses a problem in its dissipation. For power generation in particular, more than half the power has to be lost as low grade heat in cooling towers. Combined heat and power (CHP), where the waste heat from power generation is circulated round districts as domestic heating are very sensible systems, as long as the power plant is located reasonably close to living areas.

9.4b) Biomass

By tradition, the most common fuel for domestic heating is biomass – wood, crop waste, peat, dung. The fuel is locally sourced and needed in only small quantities. This may well continue in the future in rural communities, but urban dwellers will have difficulty in obtaining supplies, and so must look elsewhere.

9.4c) Solar Thermal Heating

Solar energy can be used for domestic heating, but as we are looking for low grade heat, it is solar thermal panels which simply warm water and not PV panels that are required. This is established technology in warm climates, but even in cooler climates it can be effective. It is very simple with very low capital cost. Very many houses in the Middle East have an oil drum on the roof, connected to a vertical black panel with water inside, connected top and bottom to the oil drum. By natural circulation the water in the drum heats up and serves as a preheater for the warm water system of the house. Figure 9.9 is a commercial example of these simple domestic water heaters.

Figure 9.10 Modern Thermal Solar Panels
Although dwellings in the Middle East have simple heat collectors cobbled together from a simple frame and an oil drum, the commercial equivalent shown here is much smarter, with improved collecting frame, better situated head tank, and properly lagged piping. If the water is pumped, and the water storage is indoor, the installation consists simply of roof panels, which look very similar to the voltaic cells.

China has by far the greatest amount of solar thermal heating, claiming more than 50% of the global installed capacity. It is a very old tradition figure 9.10 shows a traditional Tibetan tea seller, heating his kettle by directing the sun's rays from two metal sheets.

9.4d) Heat Pumps

Heat pumps are a more efficient way of producing low grade heat – of the order of 20 − 40 °C , from electrical power than by direct electrical heating – about one third of a kW is required to produce one kW of low grade heat. If the heat pump is just replacing the direct burning of a fuel such as oil or gas

Figure 9.11 Thermal Solar Panels of a Tibetan Tea Vendor
On the steps of the Lhasa Potala Palace, a traditional tea vendor provides hot tea, supplied from his double concave–mirrored, mobile thermal solar heating device, with the target being the tea kettle.

there is no real saving, as the creation of a half Kw of electricity requires the burning of one kW of fuel because of the waste heat associated with power generation. In short, if the heat pump is operating on electricity obtained from fossil fuels, there is technically little to be gained over the direct burning of the fossil fuel, although artificial energy tariffs may make it attractive to individual customers. If the power has been generated from renewable or nuclear fuel, then the heat pump has become the method of replacing domestic fossil fuel use by non–carbon fuel, and as such it is an important technology for reducing CO_2 from the domestic sector.

Installing a heat pump now is simply playing a game with the electrical tariff settings, but in future it will be the essential link between CO_2–free power generation and domestic energy use. Since it takes time to change, the more heat pumps installed now, the further ahead we will be when non–fossil fuels power feed the grid.

9.5) Conclusions

This chapter describes the range of possibilities for alternative energy systems which are available to replace fossil fuels. They include all the likely possibilities and mention some of the less likely ones. There may be other suggestions around, but they are so unlikely to be developed that they are not mentioned here.

What is evident is that there is no easy replacement for the large scale high availability generating sources that fossil fuel provides, The nearest to this is the use of nuclear fuels, and this does not engender enthusiasm in the population at large..

Of this list of possibilities, it is necessary to select those for development into national policies. The policies need

– integrating with existing installed plant;

– to meet targets; guard against undue risk;

– to be costed and represent a reasonably economic proposition;

– to be acceptable to the public at large.

The selection, costing and financing of these new energy sources will be the subject of the next chapter.

References

Dobbie T, *At the Core of an Industry*, The Chemical engineer, 844, Oct 2011, pg 34, (www.tcetoday.com)

Taylor, R., *Fueling the coming Era,* The Chemical engineer, 844, Oct 2011, pg 41. (www.tcetoday.com)

Further Reading

Committee on Climate Change(UK), *renewable energy review,* 2011 (www.theccc.org.uk)

IPCC, 2011 IPCC *Special Report on Renewable Energy Sources and Climate Change Mitigation:* Arvizu, D., T. Bruckner, H. Chum, O. Edenhofer, S. Estefen, A. Faaij, M. Fischedick, G. Hansen, G. Hiriart, O. Hohmeyer, K. G. T. Hollands, J. Huckerby, S. Kadner, A. Killingtveit, A. Kumar, A. Lewis, O. Lucon, P. Matschoss, L. Maurice, M. Mirza, C. Mitchell, W. Moomaw, J. Moreira, L. J. Nilsson, J. Nyboer, R. Pichs–Madruga, J. Sathaye, J. Sawin, R. Schaeffer, T. Schei, S. Schlomer, K. Seyboth, R. Sims, G. Sinden, Y. Sokona, C. von Stechow, J. Steckel, A. Verbruggen, R. Wiser, F. Yamba, T. Zwickel,[O. Edenhofer, R. Pichs–Madruga, Y. Sokona, K. Seyboth, P. Matschoss, S. Kadner, T. Zwickel, P. Eickemeier, G. Hansen, S. Schlomer, C. von Stechow (eds)], Cambridge University Press, Cambridge, United Kingdom and New York, NY, USA (www.ipcc.ch)

Hargreaves, R, (2012) , *Thorium: Energy cheaper than coal,* Hanover, NH 03755 USA (See Amazon)

IPCC, *Special Report on Carbon Dioxide Capture and Storage,*
Bert Metz, Ogunlade Davidson, Heleen de Coninck, Manuela Loos,   Leo Meyer
Prepared by Working Group III of the Intergovernmental Panel, 2005

Mackay, D. J. C., *Sustainable Energy – without the hot air*, 2012.
http://www.withouthotair.com/

Chapter 9 Sustainable Energy Alternatives

Chapter 10   Developing Energy Policies

10.1) Some Technical Consideration in Policy Making

It is a useful start to look at the development of future energy policies aimed at reducing dependence on fossil fuels from a purely technical point of view using basic principles of power generation and heat transfer. The following sections look at the points fundamental to any future policy.

10.1a) Saving and Lifestyle

We are a very spoilt society when it comes to energy use. Much of the energy is used for comfort and pleasure, which should be re-assessed when problems of sustainability of the planet are involved.

Domestic heating and heat loss can be managed to reduce consumption. Living in unnecessarily large premises is a direct waste of energy. Individuals having their own transport is not a sensible use of energy. The concept of traveling long distances simply because it is holiday time is also over–consumptive. Based on logic we should live in smaller areas with more people living together; have no car and not travel for vacation. This is an unattractive prospect as it goes against most peoples' ambitions, but it may become necessary if energy supplies become crucial.   This would probably occur naturally if energy costs rose many–fold. It could hardly be legislated for by a democratically elected government.

10.1b) Heat and Power

Electricity looks like being the primary energy system. When electricity is generated by heat with steam turbines, 55 – 65% of the energy is unused and normally is dissipated as waste heat to the environment in cooling towers. This is a loss of perfectly good low–grade heat, and as long as it is reasonably close to centers of population, it can be used for all domestic heating purposes. This means that new power plants should be 'Combined Heat and Power' (CHP), situated near urban areas and the plant size determined more by the domestic heating demand than by desired generating capacity. We would be looking at many small power generation plants in place of a few very large ones.

Nuclear power is technically the most sensible route forward. By having small nuclear plants near population centers that is the best of both worlds, as the nuclear waste heat becomes the domestic heat supply. This would suggest the need for very many small nuclear plants and a big change in the public's attitude to nuclear power.

10.1c) Low grade Heating

As explained at the beginning of the last chapter, high grade energy such as electricity is wasted if used on low temperature duties. Re–using energy streams at lower temperatures is established industrial practice, but it is more difficult to find such applications in non–industrial situations. However, at least the use of heat pumps does make better use of high grade electricity for low grade heating and it should be accepted practice that whenever electricity is used for such heating, it is to be done by a heat pump installation.

10.1d) Overcoming Intermittence of Renewable Power Generation

Wind is probably the most attractive renewable energy source, particularly where there is not a great deal of sunshine available. With its very low availability (intermittency) it does require backing up with large amounts of auxiliary quick start–up generation or some form of storage. Wind generation can usefully be used as a minor contributor, but once a significant proportion of power is generated by wind, there has to be careful provision made for its intermittency.

Offshore wind is equivalent to onshore wind but it is more expensive. If onshore wind has technical problems, then offshore wind has the same problems but is, in addition, more expensive.

The installed capacities of wind generators and solar panels are quoted in MW. This is the name–plate capacity. Their productivity depends upon the actual wind and sun conditions. Rarely will the name–plate capacity be achieved. The ratio between mean annual production and the name–plate capacity is the 'availability' or 'annual capacity factor' or 'load factor' and this is of the order of 25%– 30% for wind and 10% for solar panels. That is, between 4 and 10 times the nameplate capacity must be installed to provide the required annual power. The variations are wild, as shown by figure 10.1.

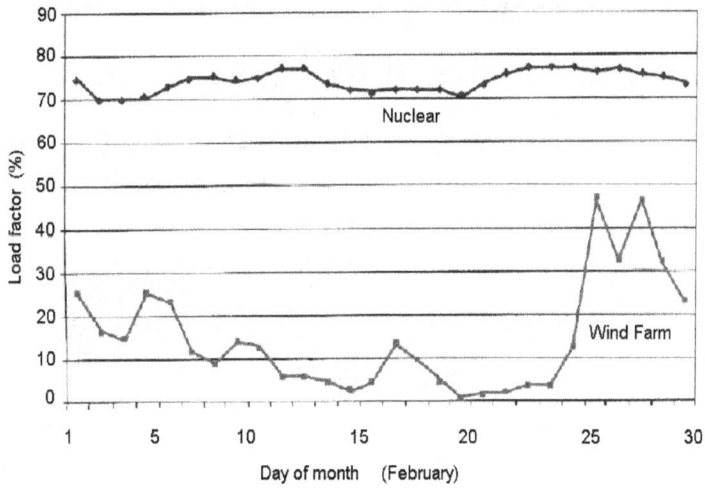

Figure 10.1 Intermittency: Daily UK wind power generation in Feb 2011
   A   major problem with all renewables is their irregular generation pattern compared to traditional methods of power generation. What makes it worse is that very large geographical areas follow similar patterns, so when the wind or solar generation is low, it is often low countrywide.

This figure summarizes the total wind generated power, each day, for February 2010 for the UK. This shows two days at 50% availability and 12 days with less than 10%.

As the number of wind generators increases, one might hope that the mean power generation would be more constant, as statistics show, the larger the number, the more stable the average. Unfortunately, our weather patterns are such that when there is wind, it is often windy over a wide area; when there is no wind, then there is no wind anywhere. This is shown by the figure 10.1, which is the cumulative output of all the wind farms in the UK in 2011.

Intermittancy

The variability between load and supply is nothing new to the generation industry. Loads have always fluctuated between day and night, and even during daytime enormous fluctuations can occur – there is the story of matching the load with generation at half–time for a cup final match, when the whole nation goes into the kitchen and switches on the kettle for a cup of tea. With intermittent generation, the problem is the same but to a larger degree.

Wind generation is playing a significant role in future in energy proposals and this intermittence problem must be satisfactorily solved.

The first rule is to have a mix of generation methods, some with stable base–loads, such as nuclear power, since this will reduce the percentage of fluctuations in generation.

The second step is to use storage to even out the power supply, storing at oversupply times, and releasing at under supply times. The favorite solution is pumped storage, with hydroelectric turbines making up the shortfall. These installations are not common because of their specific geographic requirements – in needing a situation with 2 reservoirs close together but at different levels. Other means of storage for electricity are very problematic. Battery storage is out of the question. Storage by 'chemical means ' would mean producing an energy intensive compound which could be used to release energy later. This concept is theoretically possible if we consider hydrogen to be the storage medium. Excess power can produce hydrogen by electrolysis on an intermittent basis, which could be used as transport fuel if the hydrogen fuel economy develops.

The third approach is to have extra stand–by quick–start–up generation capacity to make up the shortfall in supply. On the small scale one can talk of diesel power because of its quick start–up, or alternatively, gas–fired turbines could take up the slack. This method does require the duplication of power generation equipment, which involves extra capital cost. The intermittent use of turbines is not as efficient as a large, constantly run modern CCGT gas plant. A careful analysis can show that the CO_2 saving of the wind turbine is almost completely lost when the back–up fossil fuel generation inefficiency is included in the analysis.

The fourth possibility is to interlink power networks over a greater region. We have already said that the nature of wind generation is that a whole area, eg the whole UK, has similar wind patterns, strong winds and calm periods occur region–wide. By having a larger network – eg a whole European network, there is a much higher chance that there will be an evening out of the intermittent suppliers and that the average will be easier to handle. There are already considerable network connections across Europe, but these may have to be strengthened to cope with major wind generation variability.

Increased networks may also allow more pumped storage to be used over a wider region. In Europe, Norway and the Alps provide opportunity for this to be done, and networking would enable more countries to avail themselves of this possibility.

The fifth possibility is to control the load demand side and reduce demand when generation is reduced. This is a serious proposal and can be done with 'smart meters' which are radio controlled to switch on and off to various tariffs. By having a suitable tariff system, the less important loads could be on a circuit which can be switched off by the power supplier when he cannot match the full load. The very last resort is to accept that there might have to be the occasional power cut to balance loads. The Developed World expect 100.00% availability; the Developing World is often content with 50%

This intermittency problem has to be tackled. All the above suggestions are being employed, with the exception of hydrogen storage. As the move to wind generation is underway and many countries plan for at least 20% generation in the future, these proposals must form an integral part of all future plans.

In fact, four countries (Denmark, Portugal, Spain, Ireland), in 2010 have already been able to supply 10 to 20% of their annual electricity from wind energy, which shows that at these levels the intermittency problem of wind energy can be overcome.

10.1e) Timing

From a technical point of view, there is no reason for undue haste. The jury is out on the long–term effect of increasing CO_2. At one extreme it appears to be a manageable future, with little change as long as fossil fuel use decreases over the next century; in the other extreme, the real damage has already been done and the future will be intolerable, and inevitable. This means that in either case there is no need to make hasty changes. The advantage of moving more slowly to new energy sources is that time is available to make proper, considered decisions incorporating new technologies. With a rushed implementation, there is insufficient time to consider promising new technologies such as CSS, fuel cells and new generations of nuclear power.

10.2) Economics and Cost

10.2a) Economic Viability

Having firstly laid emphasis on getting the technology right, the next step in any investment decision is to determine whether the project is economic. This involves knowing all the monies that will be used during the whole lifetime of the project (cash flows in and out, C_{in} and C_{out}) to determine whether, at the end of the project there has been an overall loss or gain in money. This amount of money is called the Net Present Worth (NPW), and it should be positive to show the project is worth doing. If the NPW is negative, it means the project has simply lost money, which can be considered as the money lost for carrying out a project which was not financially viable, with no hope of recovering all the money put into it. The method is refined by adding a time value for the money flows, assuming it is all borrowed or repaid at the same fixed Internal Rate of Return (IRR, r) Choosing the right IRR is a matter of judgment, but once chosen, the NPW determines the financial viability of the project.

Expressed mathematically

$$NPW = \sum (C_{in,i}/(1+r)) - \sum (C_{out,i}/(1+r))$$

Where $C_{in,i}$ and $C_{out,i}$ represent total cash in and out flows for year i, and $\sum$ represents the sum of years from the beginning to the end of the project.

When choosing between alternative proposals, which is the usual situation in the case of energy projects, we can calculate the NPW for each project and choose the project with the highest NPW. If all the NPW are negative, it means it is going to cost us for the privilege of doing anything. Sometimes it is more convenient to compare different energy systems in terms of the resulting cost of energy (£/kWh – for instance). This is achieved by making NPW calculations in exactly the same way, but adjusting the revenue from the energy until the NPW becomes zero. This enables comparative tables to be drawn up for the various energy sources.

The comparison between alternatives is dependent on the cash flow patterns and the IRR chosen for the study. Choosing projects with high initial costs will be more disadvantaged by higher IRR than projects with low initial costs. The order of preference can change as the IRR changes. If we choose the inappropriate IRR then we make the wrong decisions. Inappropriate IRRs may be due to unrealistic financial demands of a capital market, or wishful thinking from investors. Socialist countries are likely to use lower IRRs than free–

markets choose, and they may therefore result in making different decisions; lower IRRs preferring heavy investments such as nuclear, instead of gas for power generation for example.

The IRR is determined from a consideration of the loan cost of risk–free capital, increased to cater for the different degrees of risk for each project, based on expert advice. Using high IRRs just because one hopes the project might make that much money only confuses the picture.

A special economic study has been commissioned by the UK government to determine appropriate costs of capital (IRR) to be expected for future energy supply projects (Oxera Consulting (2011)). This study takes into account the different technological risks of the various technologies and the expected money market. This rate of return varied between 7% and 15%, depending on the maturity of the technologies involved. Compared with reported achieved IRRs for energy projects of the order of 2 – 3% does make the Oxera estimates very high, but they are the best estimates for the cost of capital in an investment market. Engineering has to compete with retail, media, communications, entertainment and sport investment opportunities. There is only one money market.

 Neglecting to take the financial viability into consideration has serious consequences. Firstly, the project will need funding which it will never be able to repay. It will be robbing other projects, or increasing National debts with no hope of repayment. Secondly, to choose any project which is not NPW–preferred will be less economically viable than one which other companies or countries can choose. This means there will be immediate competitive disadvantages eg energy costs will be lower in other countries. Simply electing to have a disadvantage in a very competitive world is not good policy.

Competing as a Decision Criteria with the economic analysis in the decision–making is the adherence to 'policy 'and ' Public Preference'. Both these may play a part, but by overriding the most economic choice there is the immediate prospect of losing competitiveness. No more clearly can this be identified than by looking at Europe in its bid to take the lead in being 'greenest'. Policies are drawn up, economic considerations are left behind, and Europe is now fighting its high national debts and is worried about it competitiveness. By taking the eye off the financial viability ball, trouble lies ahead.

10.2b) Costs

In order to formulate a policy for future power generation, the UK government contracted an engineering company Mott McDonald to survey the whole potential processes that are available for power generation which do not involve the release of CO_2.

The Mott McDonald report (2011) has done a very thorough job at looking at over 30 non–CO_2 generating technologies; comparing their capital, operating and total power costs for each, both for the present day and for the future. Future costs are important because some newer technologies are expected to have reduced costs because of 'learning', whereas more mature technologies will gain less. All renewable energy costs are expected to fall in real terms in the future.

Table 10.1 lists the levelized costs for the different technologies, where levelized means they have included their availability (sometimes called the 'annual capacity factor'), which varies from 10% for photovoltaic cells to 80% for nuclear power plants. These tables show onshore wind to be amongst the more cost–effective technologies, together with nuclear power and CCGT gas. The next tranche of technologies include off–shore wind, and solar photovoltaic cells. Coal with CCS technology is more expensive. The other technologies have a wide scatter of costs; some would be economic in some circumstances, but none can be realized on a large scale. Run of River, (turbines in flowing rivers), waves, barrages and the various forms of bio–waste processes on an individual basis might be economic on a small scale where conditions are favorable, but for a countrywide energy policy they cannot be significant.

These costs do not include the cost of the intermittency of the wind and solar generation. There needs to be proposals made and costed on how to cope with this problem – and these costs added to the cost of these technologies shown in the table before proper comparisons can be made if generation of significant quantities from intermittent sources are to be part of the mix.

Table 10.1
Cost Estimates for Various Renewable Power Generation Technologies
(Mott McDonald 2011)

Technology	Cap Cost 2011 £,000/kW	Power Cost 2011 £/MWh	Power cost 2024 £/MWh
Onshore wind1.5	85	60	Offshore wind
3.1	160	80	
Wave+ tidal	3.8	350	150
Run of River, hydro	2.3	70	50
Solar PV	2.9	350	75
Nuclear	3.5	95	60
Coal CCS	2.7	150	120
CCGT CCS1	100	110	Biofuel
2.5 – 5.0	60–160	60–100	
Geothermal	4.7	160	80

Notes:
1) Levelised costs include the load factors and discount rates appropriate to the estimated risk
2) No cost is included for handling the intermittence of the renewables

10.3) Political Encouragement – making it happen

We have a global problem, requiring a global solution. There is no point in individual countries doing their own thing. International conferences, leading to international policies and international agreements have to be the way forward.

The first sign of international concern of the way mankind was using up the planet's resources was the meeting of the ' Club of Rome – limits to growth" in 1968 which was a symposium to discuss sustainable futures. Warnings were given regarding the unfettered consumption of petroleum reserves and raw materials in general. Probably it was here where the concepts of recycling were borne. The attendees of the symposium were unusual thinkers, with very different attitudes from those prevailing at the time in the developed world. Their messages were not really taken seriously; they remained an alternative view.

In 1987. due to concern over the disappearing ozone layer above the Antarctic caused by reaction with CFCs, a meeting in Montreal was organized in which it was agreed that those CFCs which were thought to be creating the problem should be removed from commerce, (the Montreal Protocol) This was very effective and now those CFCs are no longer used. Following the ban, CFC concentrations in the atmosphere still increased, presumably due to materials already in circulation, but since 2000 they have stabilized. Meanwhile concern over the ozone layer has declined. Since then CFCs have been associated with global warming. That they are no longer increasing is very good news.

Probably the next important international cooperation came about with the formation of the International Panel on Climate Change – the IPCC – in 1991. This is an international panel of scientists of a wide range of disciplines to discuss the effects of GHGs on global warming and general climate change resulting from mankind modifying the atmospheric composition. This committee has produced four Assessment Reports – 1990, 1995, 2001 and 2007 – of about 1000 pages each, summarizing the science associated with global warming. One of these sections, *the Physical Basis for Global Warming* is a thorough study of all the research papers and measurements associated with the evidence for global warming. The work is very thorough and earlier chapters of this book are not at variance with these reports in any factual way.

The remaining sections of the assessment reports are much less quantitative, and can only hint at potential problems that could occur with increases in CO_2. The overall result is that the IPCC reports always present a very gloomy view of the future.

One of the guiding principles of the panel is that it summarizes the present science. It is not predicting the future, which depends upon decisions which are the realm of the politicians. But being so carefully apolitical, it loses some ability to discuss the magnitude of future problems and this leads to the media filling this gap with horror stories of the future.

There is no doubt that the IPCC has greatly improved cooperation between scientists and has been able to carry out expensive combined experimental projects. Their conclusions are always hemmed in with gloomy scenarios, as their objective is to warn politicians and decision–makers of potential future problems. The result of these reports has been to ensure that politicians have held a series of international conferences to determine what actions they should take to prevent future catastrophes.

There has been a string of international conferences where an agreed approach between nations is the aim of each conference – Kyoto, Helsinki, Copenhagen, Mexico. At the first, and probably the most important conference, – Kyoto – many countries agreed to curb their future CO_2 emissions. Some countries would not agree – for instance the US, the major emitter, and the developing countries who did not emit much, but expected to in the future. This series of meetings has resulted in about two thirds of the countries of the world having now defined national energy policies in one form or another. It is only the least developed third that has not yet formed energy policies.

The US, with President Bush, gave the impression of complete denial of any problem, with no desire to agree to any restrictions. Though this looked like a selfish response by a president who could not grasp complex matters, in fact it really was the result of advice from the best scientists in the world, including the highly respected Revelle, who advised that the CO_2 problem was important to keep under investigation but it did not merit a wholesale change in economic thinking. This advice may well turn out to have been very wise.

Over a series of meetings, the developing countries have made some agreements in return for agreements from the developed nations on financial help towards establishing carbon–free economies.

The result of these meetings is that politicians return to their governments, some fired up with the desire to be seen as caring guardians of our future, others more cautious. Then come regional and national meetings which then decide how each region should best contribute to the solution.

Each government in the EU and each party in each government competes to be the most green and caring; U.S announces initiatives to slow CO_2 emissions by 50 %, the EU by 80%; and China stays well back, quietly developing its own agenda.

10.4) Financing

The politicians decide their targets, and policy to meet these targets, after considering their own philosophy, public opinion and international agreements. These policies translate to projects that are put out to tender to the private sector. These projects are rarely economic, and so the private sector requires incentives and subsidies to be willing to cooperate. There may be minimum energy price guarantees or direct subsidies.

The whole procedure is a nightmare: Too much give–away and there is public outcry of handing bonanzas to large companies: Too little, and no company will even bid for the project. There is continual criticism that the government is buying its policies at too high a price, getting senselessly uneconomic projects off the ground, ignoring market forces, at very high costs to the community.

Paying for this additional cost could come from government by increasing the government borrowing. It could be rounded up and recovered by increasing all energy costs by adding to all utility bills an 'environmental levy'. In which case those industries using energy are badly hit and are less competitive on a world–wide market. In any case somebody has to pay when uneconomic projects go ahead. Many a Chancellor in charge of government finance has warned that forcing ahead 'green' energy policies which are not economic are bad for a country's economy and competitiveness.

10.4a) Two economic systems

To make changes to the energy supply for a nation requires enormous capital for building the new facilities. This capital has to be found before building can occur. The world has two different economic systems for achieving this. The 'capitalist' world relies in attracting money from individual people or companies by showing that by using their capital they can provide as good a return as any other use of their capital. This is expected to vary from 7% to 15% depending on the risks involved. Government has to persuade asset–rich individuals that to contribute to the change in energy production is a worthwhile activity for them. Persuasion is the only tool, and this can take the form of subsidies and tax rebates in the hope of attracting their funds.

The second economic world system is socialist, where the community is more important than the individual, and once the community decides the right direction, then the funds can be allocated from all the other activities of that country to realize that policy. Analysis of the economics of any project is still done, using for instance NPW, but the cost of capital (IRR) is not as high as a individual investor would be hoping for in the capitalist economy. Risks become a risk to the state, and not a reason to require higher rates of return.

For major investment decisions in the energy sector, this second method does seem to have the edge. The next section describes some of the inducements that have been proposed to accept agreed policies.

10.4b) Carbon credits and Carbon off–setting

The Kyoto agreement in 1997 was a remarkable achievement in getting a majority of the world's nations to take global warming seriously, and to define measures to limit anthropological changes to the planet's atmosphere. Complex discussions resulted in some agreement between the 187 countries attending to reduce CO_2 emission levels by an average of around 5.% from 1990 levels in the industrialized countries by 2012. These targets seem very cautious, but it must be remembered that up to this time the energy consumption grew with the GNP, and nations would hope to have grown by more than 20% over this time period, so to agree a target of 5% reduction is quite an achievement. However fair and well–intentioned the targets, difficulties have appeared. The UK can easily meet its targets as, by chance, there was an economic incentive to change from coal to gas – the dash for gas – in the mid 1990s. That plus 5 years of near recession mean that the UK can make the saving 5%. Germany has been nearly as lucky, in annexing its poor neighbor, Eastern Germany, just after 1990. This step reduction has made the 5% saving easier. Canada, on the other hand, with no such luck is heading for a 17% increase by 2012, and therefore wished to step back from the agreement. The US did not sign up to the treaty, for fear of it adversely affecting their economic growth. They were probably wise.

Within the discussions on CO_2 emissions reduction for the developed nations, consideration was given to developing nations and a scheme was initiated to help finance energy systems in the developing world as long as they were CO_2 emission free. This is done in an association with a system of Carbon Credits and Carbon

Off–setting in an attempt at bringing financial incentives into the reduction of CO_2 emissions into industry and commerce.

The schemes require companies to be allocated an allowance, and if this is exceeded they will be fined per ton of CO_2 emitted. It may be that the companies can make these reductions in savings or policy decisions, reducing the use of fossil fuels. However, this will not be possible for all companies – depending on the type of industry, – so they will be faced with paying fines on their excesses. The introduction of Carbon Credits enables such companies to be able to take credit from the savings made in other companies, who are able to save more than their target, by means of buying from them Carbon Credits. Any justifiable and real CO_2 emissions reduction, is awarded Carbon Credits, one credit being 1 ton of CO_2 saved from emission. Any company saving more CO_2 than it needs can sell these Carbon Credits to any company unable to make its saving to be compliant with the Kyoto protocol. Hence all companies have financial incentives to save as much as possible, and not just meet their targets. These credits can also be issued by countries proposing capital investment projects to save CO_2 emissions. When these schemes are operating and actually saving CO_2 emissions they are issued with Carbon Credits, which can be purchased by those companies needing to meet their targets, or, for marketing reasons, want to appear 'green' and claim they are well above their target savings.

Hence this Carbon Credit scheme becomes a method for the industrial countries to finance CO_2 reduction initiatives in the developing world.

A carbon trading economy is developing, with markets, agents, forward trading to provide finance up–front for projects, option schemes for minimizing risk, the use of carbon credits as an investment to buy and sell at a profit as prices change. It takes on the full characteristics of the stock market, with the possibility of complete investment failure if governments no longer legislate on the CO_2 emissions or if the supply of credits through CO_2 saving projects outstrip the demand for reduced CO_2, but with the possibility of high profits if the CO_2 rules become more stringent. Particularly significant will be the future 'carbon pricing' levels to be set by individual governments.

There is the Compliance Market, where savings are obligatory to meet agreed targets, and a Voluntary Market where companies may hold more credits than they need – for instance, many supermarkets aim at zero carbon for marketing and image reasons, – but this is not done by transferring their transport fleets to renewable fuel, but by having bought sufficient Carbon Credits to match the carbon their CO_2 release.

The positive side is that the scheme releases capital for the developing countries. The negative side is that it may become a toy for the investment world, where it has a reputation of providing complex imaginative schemes which can go very wrong. To quote investment literature: *'The carbon market promises to be a lucrative market based in London'* (MH–Carbon, 2011).

To the engineer and scientist, reading that '124 million tons CO_2 have been transacted' is rather galling; No emissions have been saved; only that quantity of CO_2 has transferred ownership by paper transactions. There must be an understanding in the political world that this is not really solving the problem, as, for instance, the UK government policy on CO_2 reduction states that carbon trading will be used only as a last resort to meet legally set targets.

10.5) Public Approval

At least in the 'democratic' half of the world, nothing can be done without the approval of the majority of the population. Large capital projects such as power generation are just the type of decision which invoke intense interest and the public demand their say.

Public approval of wind farms cannot be taken for granted. There is considerable opposition because of their dominance on the landscape, and the large numbers that are necessary. Noise is also cited as a reason for refusal. Objection to wind turbines is largely subjective, and, as more and more are installed it may well wane, particularly when the object is to 'save the planet'

Hydroelectric power is more readily approved of, with the only real dissenters being those communities being displaced by newly formed reservoirs. There is however very little opportunity for more hydro power except for smaller Run of River schemes which are useful locally, but do not contribute much nationally. Hydroelectric power can only be developed in the developing countries where untapped possibilities still remain.

Nuclear Power

The most sensible *engineering* solution for reducing our reliance on fossil fuels is to use more nuclear power stations, but unfortunately this has the most public opposition. Changing to nuclear power allows our infrastructure to remain as it is, and the changes can be made in a planned, timely fashion.

The public is very wary of nuclear power. Germany and Belgium have voted to build no more. Their fears are well–founded. Nuclear reactors are capable of melt–down and devastation of large areas and populations though radioactivity. Secondly, the disposal of the radioactive waste is not fully solved.

Of the 400 or so nuclear power stations already built over the last 60 years, there have been only 3 major accidents, – 3 Mile Island (US), Chernobyl (USSR) and Fukushima (Japan)– two involving meltdown, or near meltdown . There were no fatalities in 3 Mile Island; 6 deaths attributed to cancer known to be directly associated with Chernobyl; and probably none in Japan due to the stringent precautions taken. Actual statistics show that nuclear power is the safest form of power generation. On the other hand, displacement of populations for long periods is a real problem. It is reported that over 100,000 persons were displaced by the Fukushima accident, and it is not known how long it will be before they can return.

It is possible to generate alarming figures by manipulating statistics . For example, exposure to radiation slightly increases the risk of dying from cancer – say by 0.1%. If 20 million people live within the radioactive plume of an accident, and 30% of the population have cancer cited as their final cause of death it means that

$$20,000,000 \times 0.1/100 = 20,000$$

extra deaths might be attributable to the effect of the radiation out of the expected 6 million cancer deaths that will occur anyway.

Hence we get such headlines – 'Nuclear disaster will cause an estimated 20,000 deaths'.

The storage of waste is certainly a very frightening problem, and progress is very slow. Burying a million tons for eternity is one offered solution. Reprocessing and recycling radioactive components in fourth generation reactors, or moving away from uranium–based reactors seem better solutions to hope for.

The present public conception is that:

Nuclear power involves an unnatural, highly dangerous process with potential for catastrophe during operation, and which leaves behind a lethal legacy for future generations for ever.

Or it could be:

Nuclear power is the source of energy that powers the universe – from the magma of the Earth, to the Sun and stars. Safe fuels can and should be developed, recycling of 99% of the waste could be possible, leaving only 1% to be safely buried. It will provide adequate power for a world, which otherwise will face energy shortages; with resulting unimaginable aggression.

Both these concepts are equally true. Much publicity work is needed to create informed public opinion.

10.6) Some Examples

Now let us look at some of these future policies in detail to see how the problem is being tackled

10.6a) Europe and the UK Energy policy

In the UK the Prime Minister Blair was advised by the chief scientist that the country could reduce its CO_2 emissions from power generation by 15% in the year to 2015. Armed with the information, and at a meeting of the EU, he proposed a target saving of 15% of all CO_2 emissions – 4 times more than just power generation, and the whole community agreed this figure. Returning to the UK with the European target figures, the UK then passed a law in 2008 – the Climate Change Act – setting ambitious targets of 30% reduction by 2030 and 80% reductions in 2050.

The government then set up a Climate Change Committee (CCC), lead by a group of worthy individuals, with the remit of ensuring that the Climate Change Act and its targets were achieved. This committee has determined the future plan to 2020 in detail and then considered how to reach the 2050 targets in outline.

In determining these plans, the future energy demands were developed, assuming no reduction in economic growth over the period. This total energy curve was then used to determine a future energy policy which both meets these future energy demands and the agreed emissions trajectory. Progress in implementing the policy can be determined by comparing annual CO_2 emissions with the CO_2 trajectory, and this is reported to parliament.

The first interim target to 2020 has to be done with established technologies without long lead times. This has resulted in policies which are relatively easy to implement. Beyond this date the committee needs to be relatively conservative, and not use technologies that are as yet untested. Hence the renewable technologies considered have been limited to wind, PV panels,

248

biofuels, CSS, nuclear, with limited mention of hydrogen–based transport fuel in a 'niche' market.

The committee employed consultants to look at the renewable generation area and estimate investment, operating, and maintenance costs and to give an estimate of the state of technical development of a wide range of possibilities. The resulting survey, done by Mott McDonald (2011) is a very thorough analysis, mentioned earlier in this chapter. Table 10.1 summarizes their conclusions.

A second report commissioned by CCC was to look at the financial money markets and to estimate the interest rates that would be demanded from investments in the various technologies. These rates varied from 7 – 15% depending on the maturity of the technology (Oxera 2011).

Domestic energy consumption

An easy first step is to ensure that heat and energy losses are minimized in the national housing stock:

replacement of tungsten light bulbs by energy–saving types
insulation of walls and roofs of domestic premises
encourage domestic premises to use their south–facing roofs for
        mounting PV panels
encourage the installation of heat pumps for domestic electric
        heating
replace gas boilers by the more efficient condensing design of boiler

All these measures can be taken relatively quickly and easily; they all require encouragement from the government and this is done by generous tax allowances and grants. In short, this provides an immediate saving in CO_2 but at high national expense because of the government incentives that have been offered to the public.
Any saving can be affected by what is called a 'rebound effect'. For example, after house insulation, a householder may decide to enjoy slightly higher room temperatures. Hence only part of the expected saving is a real saving of energy, the other part being used to improve

living conditions. A second example of rebound is when a more economic vehicle is exchanged for a less economic one, and the resulting fuel economy means that the owner can afford to use it more often. Again, part of the expected saving is not realized.

It has been estimated that 10 – 30% of savings expected are consumed in such rebound effects.

Power Generation

The quickest way to achieve CO_2 reductions in power generation is to use Combined Cycle Gas Turbine generation technology (CCGT) wherever possible to replace coal fired generation because of the higher efficiency and the lower CO_2 release for equivalent energy.

Wind generation is renewable technology that has been most developed, and so is ready to be implemented. The major raft of this interim plan is therefore to expand the wind generation considerably, Onshore wind has location problems, whereas the UK has abundant sites offshore, and so a large fraction of the wind generation is planned to be offshore. The decision to emphasize offshore wind was a conscious decision for the UK to become a world leader in this technology. (whereas Germany and Denmark have opted for onshore wind technology and are intending to become world experts in this technology). To this end the government has encouraged the country's electric power suppliers to supply wind farms both on and off–shore. There is a plan for steady growth for the next 10 years.

By closing coal plants first, then oil, replacing by gas will provide considerable reduction in CO_2 even though this does not move away from fossil fuel.

The longer term plans will involve more nuclear power and the introduction of Carbon Capture and Storage (CCS) technology with coal as fuel. Surprisingly, the intermediate contribution from nuclear

Table 10.2
UK Power Generation – Past Performance and Future Targets
a) 'Name–plate' Gw   b) Annual Totals

a) Technology	1990 actual	2010 actual	2020 target	2030 target
Unabated fossil– fired generation	55	32	21	0.5
CCS (coal and gas)0	0	1	10	
Gas CCGT	0	33	38	24
Nuclear	10	10	7	23
Onshore wind0	4	16	23	Offshore wind
0	1	10	24	
Marine	0	0	1	4
*Other renewables	1	2	4	4
b) Annual				
Generation TWh/yr	280	330	350	450
CO_2 emissions from generation million ton CO_2/yr	200	150	110	20

* Other renewables include hydro, biomass geothermal and solar PV

power will be less than the present because the older stations are due to be shut down and no new ones are yet being built. In the final plan, with only 20% of the 1990 CO_2 level being produced, there will be an increase in the contribution from nuclear power. Table 10.2 shows the planned generation facilities in 2020 and 2040.

Transport

There are some simple and immediate measures recommended for transport:

add a fraction of biofuels to all transport fuels
encourage the use of smaller vehicles with better fuel   consumption
encourage the purchase of new, more efficient vehicles, in place of
        older designs
encourage the use of electric vehicles and start building an
        infrastructure of charging points

Again, these encouragements are given in the form of very costly grants   to   individuals   willing   to   go   along   with   these recommendations.

Relatively easy is the proposal to have, in 2014, 5% and, by 2020, 8–10% of liquid fuels as biofuels. At these levels the biofuels can be mixed with the fossil fuel, and the consumer is not even aware that this is being done. The next proposal is for a penetration of electric vehicles, hoping for 5% penetration in 2020. There are already electric vehicles on the market, but these are not popular because of their limited mileage between charging. To counter this, the plan is for plug–in hybrids to be a major proportion of the electric vehicle market. This means that for longer journeys, fossil fuel can be used to complete the journey when the battery charge is depleted. This will involve the use of some fossil fuels, but overall CO_2 savings will have been made. There is a small concession made to the possibility of a hydrogen economy by having some hydrogen–powered buses in the later stages of the planned policy.

Progress to date

The reported progress over the first 3–4 years is that the changes to the domestic area are on track with many houses now insulated, and with a good take up of PV panels on domestic roofs. Biofuels are also being mixed with transport fuels. Wind generation investment is moving ahead with 1.9GW onshore and 1.0 GW offshore installed by 2011. The take up of electric vehicles is less than hoped.

The CCC committee is reasonably satisfied, although it admits there has been some slippage from the target trajectory. It is also clear that matters will become much more difficult later on when the annual CO_2 decrease is planned to be higher.

Engineering Criticism of the Europe/UK Plan

The root of most engineering criticism is the subjection of economic criteria to the holding to targets. Targets are to a great degree arbitrary, chosen by politicians to show good intent, but, once set, are extremely expensive to meet. But need they be met? If Europe for instance, leads the way, and moved away completely from CO_2 fuels, this would be a mere 7% difference on the global situation, so missing a target a little is not so serious that economics should be completely sidelined.

But, politically speaking, targets are targets and must be met, and met by available technology. Government incentive schemes are developed which enable the target to be achieved, but on wholly uneconomic terms. Wind is the favored solution which has been accepted politically, though more gas and nuclear power could meet targets more economically. No wonder many engineers are not happy with published future policies.

A second source of criticism is the incorporation of technologies on time scales which are wholly unrealistic, with build–times, – which must include planning approval times – and technology development times being longer than the plans allow. Even though the easiest savings are being made first, the emission savings are slipping behind targets. The targets require an ever increasing rate of emission reduction, which depends on future measures, such as a much greater reliance on nuclear energy and vast expansions of wind farms, not yet approved. Engineers doubt that such a timetable can be achieved.

In a global economy the more active the government is on green policies, the more likely it is that the manufacturing will become globally uncompetitive. Clearly, the move away from fossil fuels must be very well thought out and planned, and the speed kept in line with the whole world, so no one country is more disadvantaged than others.

From an engineering point of view, it is disappointing that the future plans rely so much on conversion to gas, and the use of CCS with fossil fuel technology. These are not moving away from fossil fuel, which must be the ultimate aim. Also disappointing is the lack of development of nuclear energy. Technically speaking this is the only source we have that offers adequate quantities of long term energy to meet society's demands.

10.6b) The Chinese Approach

Although the West expects that the growing China will ruthlessly advance its own economy as its first priority, it does look as though they will have a very mature approach to the CO2 problem.

They have a number of advantages that the developed West do not have:
   –Firstly they are in an expansion phase and need to build new plant. They are not hamstrung by having to replace existing technology like the West is.
   –Secondly, they have control over their finance capital, and can use it where and how they wish, without having to attract investors with promises of rich and risk–free returns.
   –Thirdly they can employ a realistic discount rate on investments, which may lead to better decisions than using artificially high rates demanded by the finance markets.
   –Fourthly, their populations do not expect to be able to change

254

decisions, but leave planning to the planners. Though this may not be wholly advantageous, it does make national planning very much easier than having to go through the hoops of public inquiries. From the Chinese aspect, leaving technical decisions to some random voting from the lay population is not as good as leaving complex nationwide decisions to experts.

China is embarking on a major nuclear program, to go hand in hand with their coal–fired generation program. They have a clear plan for nuclear power in the future, which means all their future energy plans are much clearer than those for Europe. The large nuclear building program will make them leaders in nuclear technology, and they will be able to put forward the best bids for nuclear builds world–wide. They are creating new nuclear designs, such as their modular nuclear reactor to cater for a wide range of capacities by multiplying up a standard design. They are operating a trial fourth generation fast neutron nuclear reactor power plant, which uses 60% of the uranium fed (not the old 1%), and can recycle much radioactive spent fuel and has less of a problem with radioactive waste. They have a Thorium Reactor development program.

They are leaders in the use of solar thermal energy, with over 50% of the world's capture of solar thermal energy being done by China.

They are the world leaders in solar photovoltaic panel production– with over 50% of the world's production in huge manufacturing facilities.

They are taking electric vehicles seriously, and are planning a factory to produce a million small, inexpensive electric cars per year.

They have also built an experimental zero–carbon city to demonstrate that they are taking the carbon emission problem seriously.

The very fact that the government and planners have more freedom to organize their future does suggest that they will be in a good position to realize a mature CO_2–limited energy policy, – though this might mean riding roughshod over public opinion.

10.6c) The Developing World

The least developed parts of the world, mainly in Africa, have a very low energy consumption, which is mainly biomass. As this part of the world develops, it will want to use more energy, and fossil fuels will seem attractive to them. It is important from a global perspective, that these areas do not become reliant on fossil fuels, but move directly to renewable energy sources.

In these regions the hydro power plants have been least developed, and there is considerable opportunity for hydroelectric scheme development. However, such schemes are capital intensive, and cannot be undertaken by a developing community without external help. Without this help they will be attracted to go to fossil fuels such as oil, since this involves less capital outlay.

A carbon trading scheme has been developed for such a situation, where, if the developing country develops a project which can be certified as being a carbon–saving project, it can issue a number of carbon bonds – one bond being equivalent to 1 ton of CO_2 *not* entering the atmosphere. This provides the finance: The purchaser of these bonds provides the capital, and in return, he knows that he has an insurance against demands that he reduces his own carbon footprint. Or it may be that the investor has an eye to the main chance and expects legislation to come into force which will increase the value of his bonds, which he can then sell at a profit to a company that needs them to meet new regulations.

10.7) Fossil Fuel Reserves

The quantities of fossil fuels available on the planet are of relevance to the whole discussion on energy policy. It may be that there is a natural limit on the CO_2 that can be liberated because fossil fuels will run out.

World reserves of fossil fuels are of great commercial interest. The US government produces figures on the world reserves which themselves are composed from data from BP and data collected from the publication *Oil and Gas Journal*.

Reserves are defined in different ways, there are '*proven* ' reserves , '*unproven*', and '*possible*'

Proven reserves *are estimated quantities that analysis of geologic and engineering data demonstrates with reasonable certainty (90%) are recoverable under existing economic and operating conditions*

Unproven reserves *– sometimes called* **probable reserves**.*– are based on geological and/or engineering data similar to that used in estimates of proven reserves, but with a certainty of 50%*

Possible reserves *– a third catagory, which include all reserves with a certainty of more than 10%.*

This data is very difficult to give with any certainty. Discoveries are continually being made and there have been scares of depleting oil resources in 1956 and 1973, only to be proven wrong, because whenever there is a predicted shortage, prices rise, more exploration is undertaken, and more reserves are discovered or considered economic to recover. This situation must one day come to an end, as there must be an upper limit to the economically extractable reserves on the planet. 2010 was considered a milestone year, because that was the first year that more oil was consumed than actually discovered by exploration. This situation results in no one taking these reserves very seriously. However, published data on reserves is all the information we have and so must accept that these are our reserves. The *unproven reserves* particularly, are a serious attempt at indicating an upper limit on fossil fuels available to us.

Table 10.3 shows the reserves of coal, oil and gas, converted into m^3 of 'oil equivalent'. The table includes both *proven reserves* and *unproven reserves*. The resulting total gives the total fossil fuel reserves from all fuels. Data is also presented of the 2006 annual consumption of these fuels, and the table also shows the years of availability for each fuel by dividing the reserves by the annual world usage at 2006 consumption rates. It is alarming to see that the oil has a lifetime for proven reserves of 43 years; gas 61years; and coal 148 years. As processes are available for producing oil from coal, or gas from coal, it is only a question of economics and price to convert

Table 10.3
Estimates of Proven and Unproven Global Fossil Fuel Reserves

	Proven reserves m3 oil equivalent	*Unproven* reserves m3 oil equivalent	Daily consumption m3 oil 2006 equivalent/yr 'flow'	Years reserves *proven* at 2006 consumption	Years reserves *unproven* at 2006 consumption
coal	448×10^9	1263×10^9	8.3×10^6	148	417
oil	210×10^9	210×10^9	13.4×10^6	43	43
gas	67×10^9	182×10^9	3×10^6	61	167
Total	725×10^9	1655×10^9	24.7×10^6	80	183

Sources
Energy Information Administration, US,
  www.eia.gov/emeu/international/reserves.html, 2009
Centre International d'Information sur le Gaz Naturel et tous
  Hydrocarbures Gazeux (CEDIGAZ)
World oil
Oil and Gas Journal,
BP Statistical Review of World Energy, June 2008

one fuel to another. This means that we need only look at the total fossil fuel reserves, knowing that any imbalance between types can be compensated for by processing. Summing the daily total equivalent m^3 oil consumption for 2006 gives a total of 24.7 million m^3 Taking the composition of oil as 84.5 wt% carbon, and density of $800kg/m^3$ enables us to calculate a yearly fossil fuel emission rate into the atmosphere as;

$$24.7 \times 10^6 \times 365 \times 0.845 \times 0.800 = 6.1 \text{ Giga tons C in 2006}$$

(This compares adequately with the The Carbon Dioxide Information Analysis Center (CDIAC) data of 7.5 Giga tons for 2006, – see chapter 5.)

The total fossil fuel reserve using the 2006 usage rates. will last 80 years on proven figures and 183 years using unproven figures,

But energy consumption will not stay at 2006 rates! Chapter 5 has attempted to predict future planet fossil fuel use in order to determine the potential impact of the emitted CO_2, and we can use these estimates to predict more realistic consumptions of fossil fuels than just considering 2006 usage rates. The predicted rates are given on figure 5.6. Let us recap the basis for this figure, because the conclusions are going to be important:– this figure assumes that the developing world will increase at a 4% annual growth rate until it matches the developed world consumption per capita, together with a 1%/annum reduction in the developed world because of CO_2 reduction measures.

Table 10.3 gives total proven and unproven reserves beyond 2006 as 490 and 1118 Giga tons C respectively. Using the figure in chapter 5, it is clear that our consumption curve will overtake the proved reserves figure in 2040n and the unproven figures will last only up to year 2094 – which suggests we have between 30 and 80 years of reserves left.

These figures are alarming. Availability of fossil fuels is equally as serious as CO_2 emissions, and almost makes the discussion on future CO_2 levels redundant because there may be no fossil fuels left by the years 2100. Clearly all this is rather simplistic, our fossil fuel consumption curve will be a smooth transition between stages, and fossil fuels will never run out, they will just become more and more expensive. But the message is clear, for one reason or another we should learn to live without fossil fuels.

This analysis is all the more alarming if we consider that many future energy policies involve the use of CCS, which, by the time coal is being used , requires an increase in consumption by a further 40% – reducing further the number of years before reserves are use up.

The evidence for serious problems from CO_2 emissions is far from certain. The evidence for the depletion of our fossil fuel reserves is also far from certain. Together it is clear that our reliance on fossil fuel must be reduced; there will be one limit or another to their use.

10.8 Conclusions

It is very hard work to change energy technologies in an established society. In the West there is an adequate system operating which is economic and well established. To change, to reduce CO_2 emissions, requires alternatives to be built which are less economic than the established system. Finance can only be found when it is heavily subsidized and this leads to expenses for the government and for society, with no apparent return. This is particularity difficult because the objective set have been purely self–imposed. The 5% CO_2 saving of Kyoto, has moved to 80% for EU and UK targets, without there being any economic pressure to do so. All decisions become controversial, subsidies become debatable, with all decisions subject to public scrutiny. No wonder the whole seems most unsatisfactory, and it is unlikely that the self–imposed targets that have been set will be achieved.

From all reports, China has an easier problem, It needs more energy resources, it is less beholden to public opinion, and it has ample finances at low interest rates available to it. This reflects in the measures it announces, which include a lead in nuclear energy development, and a major development of electric cars.

From a technical point of view, the policies in the West are not very adventurous. There is a reliance on gas technology, because it produced less CO_2 per kWh simply because of it molecular formula, and a hope to go to carbon capture and storage (CCS), because it keeps existing fossil fuel plant and technologies intact, and a move to electric vehicle, returning to visions of the electric milk floats of the 1930s. Wind will always remain a minor item, and hydro power will

260

be used wherever it can, but it will be very limited. Technically, long–term it would be progress to see a transport system developed around H_2 generated from renewable power, and more development on the nuclear front.

These developments will only occur when the economics is right, and at present there is no economic incentive for any immediate change in the major transport propellant, and the public is set against the word 'Nuclear'.

The sting is in the tail, and the analysis in the last section on fossil fuel reserves suggest that in between 30 and 80 years time there will be desperate fossil fuel shortages. This will probably make the CO_2 problem a thing of the past, and there will be intense economic activity creating new energy forms – most probably including nuclear and H_2, which will take over from the depleted fossil fuels. At such a time, prices will be high enough that subsidies and incentives will not be necessary, and the private sector will be grasping the opportunity to replace fossil fuels, because there will be sufficient profit in such activity.

References

Committee on Climate Change (UK), M*eeting Carbon Budgets – 3ʳᵈ progress report to Parliament* 2011 (www.theccc.org.uk)

Mott McDonald, '*Cost of low carbon generation technologies*' – May, 2011,committee on Climate change(UK) , (www.theccc.org.uk)

Oxera, '*Discount Rates for low carbon and renewable generation technologies',* – prepared for the committee on climate change (UK), April 2011, (www.theccc.org.uk)

Further Reading

Committee on Climate Change (UK), 'Renewable Energy Review, – *Renewable electricity generation scenarios*',   2011, (www.theccc.org.uk)

MacKay, D. J. C.. '*Sustainable Energy—without the hot air*'. Cambridge: UIT. (2008) http://www.withouthotair.com.

MH–Carbon, www.mhcarbon.com, 'An introduction to carbon Credits', 2011

Chapter 11   Some Overall Conclusions

Having adhered to a scientific analysis of all the various factors involved in the global warming debate, it is now time to draw things together. Now is the time for introducing a little judgment and subjective analysis, in order to draw useful conclusions. This chapter contains guesses and hunches coming out of the scientific analysis of the preceding chapters. They are not intended as statements of facts but suggestions for further verification and study.

Here is a list of points which seem to stand out from the previous 10 chapters.

11.1) CO_2 and Global Warming

– What is known, What is not known

It is very certain that the CO_2 in the atmosphere is increasing because of fossil fuel burning, and the levels are higher than the planet has seen for thousands of years. There is no doubt that the levels will keep on rising. There is also sufficient evidence from physical radiative theory, that the increases in CO_2 will cause a warmer planet. However, the CO_2 alone is not likely to produce large temperature rises. For the temperature rises to be alarming, more than just the CO_2 changes must be involved. Planet temperatures are rising, and a number of different methods of measurement all show that this is the case. However, the temperature rise is not alarming and a relatively simple analysis suggests that this temperature rise will not cause great difficulty. Extensive scientific analysis, using complex General Circulation Models suggests that the small CO_2 temperature rise will induce more changes or feedbacks which, together, will produce unacceptably high warming. These predictions are not precise, and the models produce a wide range of results –from a hardly significant 2°C rise to a serious 5°C or more. There needs to be a much more precise definition of what will actually happen before major decisions can be made.

The General Circulation Models do not appear to be able to narrow down this range of results and there is no easy way of analyzing the future in other ways. We are left with the likelihood that the future temperatures will be tolerable, but there is a small chance that feedbacks will make the future much more problematic.

Looking back at past temperatures and CO_2 levels on the planet is potentially one way of resolving this dilemma. Paleographic studies show that temperatures have always fluctuated, as does CO_2, but the CO_2 increases follow the temperature, and so CO_2 is not the cause of temperature rise. Planet temperatures do change cyclically by about 5– 8 °C on a 10,000 year cycle, which results in the appearance of ice ages. These ice ages could be construed as demonstrating that the planet temperature is highly sensitive to small changes. Ice ages are thought to be induced on cycles triggered off by the planet orbital changes, and these changes are magnified by the changing albedo of the polar regions. In fact the growth and decline of the ice caps can be satisfactorily explained in terms of changing albedo and ice cap areas. There is no need to invoke any other potential feedbacks to explain these planetary oscillations.

The sea appears to absorb more CO_2 than would be expected from the active sea depth and the chemical and physical equilibria involved. One explanation for this might be the dissolution of deposited $CaCO_3$ which increases the alkalinity and therefore the capacity of the sea to absorb more CO_2. This acts as a buffer and prevents the sea composition changing. If this improved absorption by the sea were to come to an end, then there would be great concern over the effect the increased pH and decreased CO_3'' levels would have on the sea biosystems. This is probably the major concern posed by the increasing CO_2.levels on the planet.

11.2) The Need to Change

It could be argued that as the evidence for catastrophic changes from anthropogenic CO_2 are far from certain there is no need to change our fossil fuel energy policy. But there are a number of reasons for changing:–

1) There is a possibility that the generated CO_2 will cause climate problems

2) There is a limit to the fossil fuel reserves and we should be learning to live with less reliance on this single fuel

3) The supply of this limited quantity of fuel will cause major political problems and instability. There will be attempts to control areas of production, and price variations induced by the producers will destabilize the economies of the rest of the world.

These reasons are good grounds for looking seriously for alternative energy supplies to reduce our present dependence on fossil fuels.

11.3) No Undue Haste

There is an urgency around global warming which has been promoted by emotional hype rather than scientific reasoning, saying that immediate action is necessary to save the planet. Pro–active politicians demand strong policies, and embark on ambitious programs for energy substitution. The evidence is not there for an urgent response. By insisting on quick action, unsuitable and uneconomic energy policies will be embarked upon, which will not be the best long–term decision and not allow development of better alternatives to occur.

There needs to be steady progress. There is little advantage for any country to try to be a leader in carbon reduction as it is likely to be damaging to its manufacturing base.

11.4) Heeding Economics

All alternatives are less economic than the established fossil fuels processes. To look at alternatives does mean that the normal economic criteria must be subjugated to some principle of policy. Economic criteria must therefore be relaxed to allow any changes to be considered, but this relaxation should not mean complete rejection of economic principles. Only reasonably economic suggestions should be supported. The idea of accepting a wholly uneconomic proposal, simply to make progress on a CO_2 target is not advisable as it will only lead to difficulties in the future. There is little point in borrowing money, which *must* be repaid by future generations, simply to tackle a problem which *might* affect future generations. We are simply exchanging one problem for another for them.

Economics is not just about profit generation. Economic analysis is a means of ensuring that a society can afford the plans that it wishes to make.

11.5) Setting Unattainable Targets

Part of the politics of the climate change is to create long–term plans for CO_2 reductions and then to adhere to them, discussing legally binding targets and fines for non–achievement. It is good that there are plans, but they should, from the start be achievable. The UK plans of a 3% per year drop, and 80% reduction from 1990 levels by 2040 appear unachievable, when the only progress 7 years into the plan, can be attributed to a one–off 'dash for gas' giving a step reduction in CO_2 emissions from power generation. Other moves and policies do not seem to be effective, mainly because they are too small in comparison to our energy demands.. It will be very depressing to see none of these targets being achieved, it is better to be realistic from the start.

11.6) Lifestyle Changes

A major contribution to reducing energy consumption would seem to be a return to lifestyles of 50 – 100 years ago. Has there been a distinct improvement in the spiritual well–being of individuals over this time, with energy consumptions having increased 4 fold? More comfortable living, much more travel and material consumption have occurred through changes in lifestyle but has this improved our life experience, or is it just a trend in fashion? An alternative lifestyle could be envisaged being equally satisfying without the consumptive habits we have evolved. How populations can be convinced that high energy consumption is not an essential part of human life will be difficult to achieve. It may well be that the only way will be by massive price increases for energy, which may occur anyway.

11.7) The CCS Tragedy

It is alarming to see best UK plans for removing CO_2 emissions still require around 50% of our energy to be provided by fossil fuels in 40 years hence. In order to maintain CO_2 reduction targets, this 50% is planned be fitted with Carbon Capture and Storage (CCS). It looks as though fossil fuels must stay with us, but we can prevent the CO_2 entering the atmosphere. This suggestion is the worst of all worlds. The CCS will cost up to 40% of the energy generated and electricity generation will again give only a 40% generation efficiency. Allowing 10% transmission losses will mean that to produce 1 Kwh of energy will have required fossil fuel energy of 4Kwh to be consumed. To make matters worse, this might be for domestic heating, – to save us from wearing warmer clothes!

Fossil fuels are a rich resource of raw material for our society. Plastics, constructional materials, fertilizers, pharmaceuticals. all come from fossil fuels which are chemicals, partially ready to be made into these products. To be using these chemicals simply to extract their energy and at 25% efficiency is the height of disregard for future generations. By accepting CSS, one is going down this route. If CCS is not considered to be socially acceptable behavior between generations, then more effort would be put into funding development into alternative energies.

11.8) Nuclear Power

As much as it is unpopular, it seems inevitable that the world must rely to a very large extent on nuclear power in the future. That it is so difficult to move away from fossil fuels indicates that there must be more emphasis on the nuclear option. It has an acceptable safety record, and there are new designs which should further improve these aspects. New fuels, new processes and improvements in fuel reprocessing are all open to development, and will make nuclear power more acceptable. Admittedly, there is the background fear of some nuclear catastrophe with radiation poisoning or displacing thousands of people. This possibility must be compared with the world unrest that could occur politically if, through energy shortages, a crisis really took hold in an energy–rationed society.

11.9) Transport Fuels

New transport fuels seem to present a brighter picture than most other energy supply scenarios. Although the use of hydrocarbon liquid fuels have great advantages in storage, transport, and transfer into vehicles, it does seem that alternatives are not far away. Electric vehicles are just becoming acceptable, although long journeys and long charging times will always be problematic. Hybrid vehicles are already accepted by the market, and although they do not involve much reduction in carbon emissions, they are a move towards electric propulsion, and the plug–in hybrids will be more acceptable in the market than the pure electric vehicles.

The really exciting development will be hydrogen fuel cell vehicles, and these are already well developed. With a better efficiency and pollution–free fuel cells the future looks very promising. The development does not seem far off, as it is no great step to replace the internal combustion engine of the Hybrid with a fuel cell. Transport and distribution of the hydrogen fuel does not seem a great headache and small modular electrolytic hydrogen generators and compression units at filling stations might be possible.

Of course all these new possibilities do rely upon the generation of more electric power, and this must be done without fossil fuels. All roads seem to lead to more nuclear power.

11.10) Newer Energy Sources

There do not seem to be any inspiring new possibilities for renewable or non–fossil energy provision in the pipeline. Wave energy collection, with its 'nodding ducks' does seem to be too disperse and will probably require too much maintenance to be a useful provider of electricity large scale. Nuclear fusion is a process as far from application now as it was 50 years ago, and though it promises unlimited energy, we are not needing unlimited energy. Nuclear fission power would provide all that is necessary.

Hydroelectric power is of course ideal, but virtually all possibilities are used up. Estuary barrages are as fine an idea as hydroelectric power, but they do rely on finding the estuaries. Where they can be installed economically they should be, but there is a limited number of estuaries that can be developed.

The world is desperately short of a publicly acceptable, non–fossil fuel process for the large–scale generation of electricity.

11.11) Some Personal Observations

Having spent two years looking in detail at the whole question of global warming, starting with a slightly skeptical impression, but wanting to investigate the science involved by going back to the original papers, my overall impression now is that global warming is not one of our most serious problems and there is no great need for immediate action. The simple evidence is that there will be a rise of about 2°C, one degree of which we have already experienced and which was fairly difficult to notice. The excess CO_2 will reside in the sea and change its pH, but not beyond natural past variations. The evidence on increased weather patterns does not stand up to statistical investigation, except for a small rise in precipitation, corresponding to the rise in water vapor pressures at the higher temperatures.

Suggestion of planet feedbacks which will magnify the little warming of the CO_2 itself are difficult to justify:–

> *Water vapor* has a calculated feedback which is not excessive, which may even be reduced by increasing cloud cover. Attempts at measurement have not observed any water vapor feedback.

> *Ice cap* feedback – this is an important feedback during ice ages where the amount of ice is considerable, but with the present amount of ice the effect is negligible and will get less as the icecaps shrink.

> *Thermal lags* due to ocean heat capacity are difficult to imagine and fly against the measured evidence that sea water temperatures rise very similarly to mean global temperatures,

> *Aerosols* hiding the real global warming are difficult to believe, as aerosols are not observable or measurable.

> *Methane release* from permafrost, must have occurred between ice ages, and we have the resulting methane measurements – they are well below today's values.

Without these feedbacks, the global warming, calculated from infrared absorption data is not critical, or demanding of immediate action.

The major worries and suggestions of catastrophe arise from the use of General Circulation Models, developed for weather forecasting, but now used to predict weather patterns decades ahead. These models are tuned, and very accurate for 5–day forecasts, but are not at all tuned to increases in CO_2 and the resulting effects. The values of the parameters in these models have not been properly justified by tuning. It is up to the meteorological and climatology communities to justify their claims of feedback and severe weather patterns without hiding behind extensive models and huge computers.

It is not usual in global warming texts to even hint at there being any advantage to a mild degree of global warming or CO_2 increase.

But it is known that the rate of growth of plants on land is increased by increasing CO_2 concentration in the atmosphere. In fact some tomato growers already feed CO_2 to their greenhouses to improve their yield – increasing agricultural production will be one of the needs of the future. A mild degree of global warming will make some of the frozen North (such as Siberia and Canada) more habitable and amenable to agriculture. This will also relieve food supply problems as populations increase. A little warming will increase the amount of rainfall, and this will alleviate water supply problems which are bound to arise as water becomes a scare resource. The picture is not completely one–sided.

Of course it is not good to increase the CO_2 in the atmosphere well beyond values that the planet has experienced in the last few million years. Likewise, it is not good to have populations on the planet of 7 – 9 billion, well beyond anything the planet has seen. It is not good to be cultivating the whole surface of the planet with monoculture to feed the population. It not good to be genetically modifying plant and animal life to increase food productivity. It is not good to use nuclear energy for our energy requirements, producing radioactive waste with no final solution. We are having to undertake all these ill–advised activities because of our population growth and our human rights notions that preclude moderating population growth by any means. We have completely circumvented nature's limits to population growth, which involve disease, starvation and aggression

as natural means of control.   There is much we do which is ill–
advised, the burning of fossil fuels is just one of them.

And finally, what will the world look like a few decades from now –

*The contribution of the renewables may not have met
expectations, and the world may still be reliant on fossil
fuels. Prices will rise and poor countries will be left behind.
Eventually resources will become scarce and the stronger
nations will start occupying the sources of production on
various pretenses. Economies will stagger – all very ugly.*

Or, more optimistically,

*Though contributions by renewables helped somewhat,
development of the many possibilities with nuclear power
might have been used to generate the necessary power,
using cleaner fuels, some uranium some thorium based,
recycling of waste to a great extent, with   security and
safety risks being minimized by careful choice of fuels.*
*Transport might be all electrically power, by battery for
local transport, but with hydrogen for longer distances.
Economic growth will not suffer, and the poorer nations
will be catching up with the developed world.*

Appendix 1 – Some Basic Data

Planetary data

Diameter of the Earth	12,742 km
Surface area of the Earth	5.10×10^{14} m^2
Mass of atmosphere	5.13×10^{18} kg
Mass of seawater on planet	1.29×10^{18} tons
Solar radiation at TOA	1365 w/m2
Mean insolation reaching Earth surface	240 w/m2
Mean surface albedo	0.70
CO2 content of atmosphere at 280ppmv	2.33×10^{12}
CO2 content of oceans	1.29×10^{14} tons
Fraction of globe covered by water	70%
Ocean mean depth	3.6 km
Density of sea water at 5 °C, salinity 35g/kg	1028 kg/m^3

Units and Conversions

n	nano	10^{-9}
μ	micro	10^{-6}
m	milli	10^{-3}
k	kilo	10^{+3}
M	mega	10^{+6}
G	giga	10^{+9}
T	Tera	10^{+12}
P	Pica	10^{+15}

Energy 1 kJ = kW s = 0.2778 10^{-3} kW h

Radiation

Frequency υ s^{-1} or Hertz (Hz)
Wavelength λ μ m
wave number η cm^{-1}

$$\upsilon = \omega/2\pi = c/\lambda = c\,\eta$$

where   c = velocity of light = 2.998 x 10^{8} m/s

Stefan–Boltzmann Constant = 5.670 x 10^{-8} $W/m^2\,K^4$

Molecular Weights

Name	Formula	Molecular weight
Carbon	C	12
Hydrogen	H_2	2
Oxygen	O_2	32
Nitrogen	N_2	28
Carbon dioxide	CO_2	44
Water	H_2O	18
Calcium carbonate	$CaCO_3$	100
Calcium bicarbonate	$Ca(HCO_3)_2$	162
Methane	CH_4	16

Energy content of Various Fuels

Fuel	Heat of Combustion $(lbs/10^6\,Btu)$
Natural gas	14.8
Gasoline	13.0
Fuel oil	12.8
Coal	8.0
Hydrogen	30.0
Wood	4.4

Standard mean chemical composition of seawater at Salinity 35 g/kg

Ion	g /kg	mol /kg
Cl'	19.35	0.545
Na+	10.78	0.469
Mg^{2+}	1.283	0.0528
SO4"	2.712	0.0282
Ca^{3+}	0.412	0.0102
K+	0.399	0.0102
CO2	0.0004	0.00001
HCO3'	0.1080	0.00177
CO3"	0.0156	0.00026
OH'	0.0002	0.00001
Total	**35.1**	**1.12**

Vapour Pressure of Water

Temperature K	Temperature Deg C	Water vapor pressure millibar
243	−30	0.44
253	−20	1.13
263	−10	2.67
273	0	5.81
286	13	14.5
287	14	15.5
288	15	16.5
289	16	18.8
291	18	20.1
292	19	21.4
293	20	22.8
298	25	31.6
303	30	42.4
308	35	56.3

Appendix 1 – Some Basic Data

Appendix 2   The CO2 Neutralization Reaction

Some Thoughts on its Importance

Chapter 5 develops an absorption model for the atmospheric CO_2 which could be fitted very well by a semi–theoretical model. This model was then used to extrapolate for the duration of the fossil fuel reserves to determine the expected concentrations of CO_2 in the atmosphere and in the sea. This fitted model could be seen to be physically incorrect because it assumed the seas were totally mixed, but since the model fitted so well it could still be used for extrapolation with care, accepting that the more the extrapolation, the more error will occur, as was illustrated by the CO_2 concentration model discussed in chapter 1.

Chapter 5 suggested that there was evidence for a greater potential for CO_2 transfer into the sea than the simple absorption model predicted, and that the only way that the sea could be more absorbent of CO_2 was by increasing its alkalinity. The alkalinity of the sea is changed by the Ca^{++} content, and this is changed by the degree of re–solution of $CaCO_3$ by CO_2 in the *CO2 neutralization reaction* which is part of a set of equilibrium reactions:

$$CO_{2\text{ Gas}}$$
$$\updownarrow$$
$$CaCO_3 \quad \leftrightarrow \quad CaCO_3 + CO_{2\text{ Liq}} + H_2O \quad \leftrightarrow \quad Ca^{++} + 2HCO_3{}'$$
$$\text{solid} \qquad\qquad \text{dissolved} \qquad\qquad\qquad\qquad \text{increased alkalinity}$$

This can all be modeled, but as is all conjecture, the modeling of this suggestion is confined to an appendix to the book, which is concerned with discussing matters which are well founded. Separating it from the main text is to ensure that the message of the main text is not confused by discussion on the likelihood or not of the neutralization reaction being significantly involved.

Appendix 2   The CO2 Neutralization Reaction

This appendix develops an absorption model with only 700m depth of sea being active, but including the possibility of the CO_2 reaction with solid $CaCO_3$ beneath the seas.

The Model

Let us assume that there is an equilibrium occurring between the solid $CaCO_3$ and the CO_2 in the sea and that at 280ppm CO_2 in the atmosphere, in dynamic equilibrium with a specific CO_2 concentration in the sea. As CO_2 concentrations in the atmosphere increases, the CO_2 sea increases, and this can be followed by the calculated CO_2 partial pressure of the seawater (p*). The neutralization reaction will then occur proportionally to (p* − 280), the distance from equilibrium. Assuming a normal kinetic relationship, then the rate of the reaction R4 will be

$$R_4 = K_4 (p*-280) \qquad \text{mols } CaCO_3 \text{ /kg sea/year.}$$

One mole of $CaCO_3$ dissolved by this reaction will add 1 mole of CO_2 (DIC) and increase the alkalinity (alk) by 2 moles, because Ca^{++} has 2 positive charges. Hence our previous model described in detail in the Chapter 5 Appendix requires only 2 additional terms to the DIC and alkalinity equations : viz

new DIC content of the sea:
$$Msea(n) = Msea(n-1)+T(n) +K_4(p(n-1)*-280)$$

new alkalinity of the sea:
$$alk(n) = alk(n-1) +2\,K_4(p(n-1)*-280)$$

This model can no longer be represented by the Revelle factor, which we fitted from the equilibrium spreadsheet described in the Chapter 3 Appendix only once and produced an analytical solution for the mass balance spreadsheet. Now, each year we must evaluate the new pH and p* from the equilibrium spreadsheet with the new alkalinity, and enter the p* in the mass balance spreadsheet before proceeding to the next year.

Model Results

There are now two unknown model parameters, Kga and K4, and these can be found by fitting the Kga from the early–years profile of CO_2 and then adjusting the value of K4 by fitting in the later years. The fit is shown as figure A2.1 – it is very good, as good as the 4000m active sea model. Compare Figure A2.1 with Figure 5.3a of chapter 5.

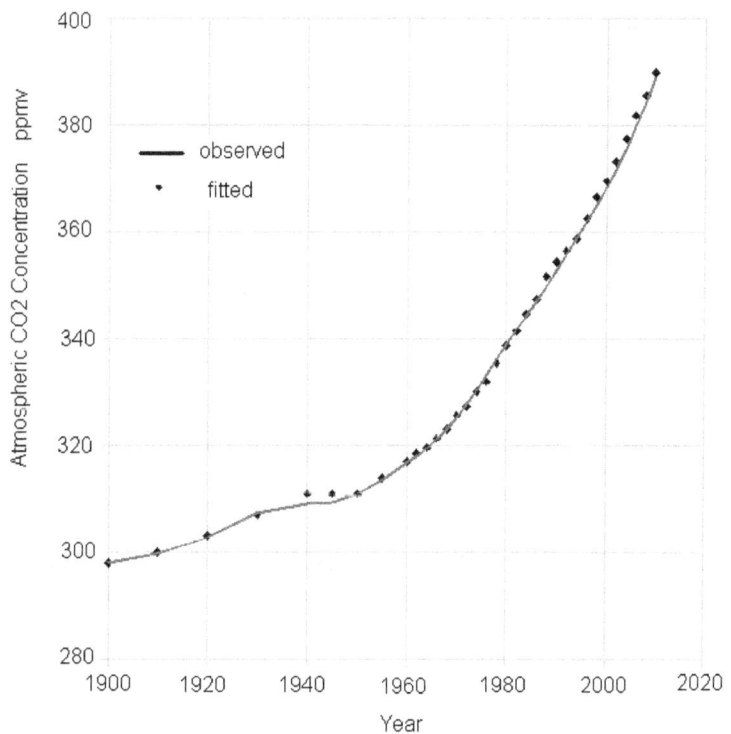

Figure A2.1
Fitted Atmospheric CO2 for 700m active Sea Depth with Neutralization Reaction

The model can be fitted to the observed data by searching for values for the 2 parameters Kga and K4. This fit is very good, which gives confidence in using it for extrapolation, and also is evidence that the suggested mechanism is plausible.

Best fitted values were Kga = 2.2 x10^8 and K4 = 2.17 x 10^{-8}.

Appendix 2   The CO2 Neutralization Reaction

Using this fitted model to predict the future is shown on Figure A2.2. It is very similar to the 4000 active depth model, but there is a slight difference in the atmospheric CO2 level predicted, – see Figure 5.7 chapter 5.

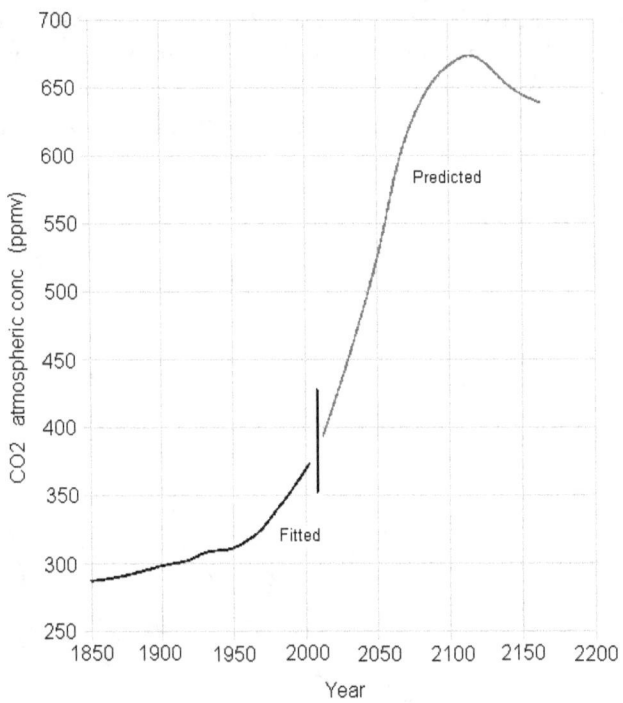

Figure A2.2   Predicted Future Atmospheric CO2 levels using the '700m active depth with neutralization reaction' model

The parameters fitted to the data in figure A2.1 were used to predict the future CO2 levels with the emissions curve developed in chapter 5. The resulting curve peaks at a somewhat higher level of 670 ppm compared with the figure 5.7 of 620ppm. The general principle is that the neutralization reaction provides a much greater capacity of the sea to absorb CO2, making the 700m active depth have an equivalent capacity to the 4000m depth used in the chapter 5 model.

Tables A2.2 and A2.3 at the end of this appendix are outputs from the spreadsheet calculation for the fitting and prediction sections of the model. These can be compared directly with the tables in chapter

5, which assumed 4000m sea depth and no neutralization reaction.

It is interesting to compare the various sea water compositions for the cases we are discussing. Table A2.1 summarized these predicted compositions and compares them – and for interest compares them also with calculated figures for rainwater.

Table A2.1
Seawater Equilibrium compositions calculated for the Past, Present and Future

Case	Year	Alkalinity mol/kg	DIC mol/kg	pH	pCO2 * ppm	CO3" mol/kg	HCO3 ' mol/kg
Rain water based on CO2 equilibria	1900	0.00E+00	1.12E–05	5.48	280	1.19E–09	3.31E+06
Rain water based on CO2 equilibria	2004	0.00E+00	1.45E–05	5.41	375	1.19E–09	3.84E–06
Standard sea	1900	2.35E–03	2.03E–03	8.19	280	3.15E–04	1.71E–03
Standard sea	2004	2.39E–03	2.10E–03	8.14	326	2.98E–04	1.79E–03
Sea, *4000m active depth* **without** CO2 neutralisation	2150	2.35E–03	2.11E–03	8.04	440	2.44E–04	1.86E–03
Sea, *700m active depth* **without** CO2 neutralisation	2150	2.35E–03	2.24E–03	7.72	≈1000	1.30E–04	2.09E–03
Sea, *700m active depth* **with** CO2 neutralization	2150	3.09E–03	2.77E–03	8.05	506	3.27E–04	2.43E–03

The figures in this table, which are based on calculating the composition of the sea from the amount of CO_2 from the air it has absorbed, suggests that the sea has changed pH by 0.05 units since 1900 to the present. The future composition depends upon the model chosen. For 2150, pH has dropped 0.15 units for the *4000m active depth* case, 0.42 units for the *700m active depth **without** neutralization case* and 0.14 units for the *700m **with** neutralization* case.

The 4000m case we know is not real; the 700m **without** neutralization gives very serious changes to sea water composition, but it does not agree with observed CO_2 absorption, but the 700m case **with** neutralization does fit measured data, suggests that the future sea water compositions will not be problematic. It is clearly very important to determine the role played by the CO_2 neutralization reaction.

The CO_2 neutralization reaction is known by oceanographers to be very important. It is part of the final solution of the CO_2 problem. It has been discussed in very many papers (Archer, 2009, Feely, 2004), but the general consensus is that the reaction will take centuries to reduce CO_2 levels. The work here indicates that the reaction is more significant now than these papers suggest, unless we can find another explanation for the improved absorption of the sea as the CO_2 concentration rises.

Another question is the quantity of $CaCO_3$ available and where it is located. Some papers suggest it is at the sea bottom, and so, because of mixing, reaction will take centuries, and others suggest it will be in surface waters. The model suggests that there will need to be 8000 giga tons of $CaCO_3$ to dissolve − six times more ton−wise than fossil fuel burnt. Some estimates suggest there is about the right amount of $CaCO_3$ available to absorb all the CO_2 from the fossil fuels eventually (Archer, 1997).

The neutralization reaction also takes an important role in the discussion on CO_2 atmospheric variations during the ice ages. Figure 8.7 shows the atmospheric CO_2 levels do oscillate as the temperatures oscillate with ice ages. This is thought to be due to the changes in sea alkalinity by more or less conversion of Ca^{++} to solid $CaCO_3$. Precipitation being caused by the biosystem activity, and dissolution being caused by high CO_2 concentrations and the neutralization reaction (Sigman, 2000).

So the neutralization reaction could be key to the whole fossil fuel CO_2 problem.

Table. A2.2 Model Fitting   700m sea depth with CO2 neutalization Model

year	c emission	arc ml btea	co2yln air	Gum co2ln air	calc ppm	Obj ppm	% btea	total cumd	p isa	Totc inisa	DIC	glob warm	Blk TA
2.2E+8													
1750	3.00E+006	6.28E+06	4.72E+06	2.35E-12	280.0	280.0	1.00	4.0000E+00	280.0	2.3165E-13	1.1100E-01	0.00	2.1700E-08
1800	8.00E+006	1.25E+07	1.69E+07	2.35E-12	280.1	280.0	57.09	7.5000E-07	280.0	2.3165E-13	2.0327E-06	0.00	2.3900E-03
1825	4.50E+007	3.45E+07	1.62E+09	2.36E-12	280.2	280.0	42.46	3.5000E-08	280.0	2.3166E-13	2.0328E-06	0.00	2.3900E-03
1850	4.60E+008	1.03E+09	6.61E+08	2.40E-12	285.0	285.0	2.09	6.0750E-09	280.3	2.3166E-13	2.0329E-06	0.03	2.3900E-03
1875	4.60E+008	7.54E+08	1.41E+09	2.41E-12	286.9	288.0	60.79	1.7490E-10	283.5	2.3192E-13	2.0353E-06	0.04	2.3903E-03
1900	5.90E+008	1.57E+09	2.28E+09	2.45E-12	291.1	291.0	34.86	3.0575E-10	284.0	2.3211E-13	2.0388E-06	0.06	2.3541E-03
1910	1.05E+009	2.40E+09	1.63E+09	2.50E-12	297.9	298.0	40.68	5.1075E-10	287.0	2.3250E-13	2.0444E-06	0.09	2.3885E-03
1920	1.10E+009	2.55E+09	2.39E+09	2.52E-12	295.9	300.0	59.51	6.1825E-10	288.2	2.327+E-13	2.0481E-06	0.10	2.3615E-03
1930	1.35E+009	2.46E+09	3.95E+09	2.54E-12	302.7	303.0	51.79	7.4075E-10	291.5	2.3300E-13	2.0521E-06	0.12	2.3651E-03
1940	1.75E+009	3.39E+09	1.88E+09	2.58E-12	307.4	307.0	38.37	8.9575E-10	292.0	2.3324E-13	2.0567E-06	0.14	2.3701E-03
1945	1.30E+009	3.53E+09	7.26E+08	2.60E-12	309.0	311.0	71.11	1.0482E-11	293.0	2.3358E-13	2.0623E-06	0.15	2.3753E-03
1950	1.16E+009	3.40E+09	2.57E+09	2.60E-12	309.5	311.0	82.93	1.1097E-11	294.0	2.3376E-13	2.0653E-06	0.15	2.3781E-03
1955	1.63E+009	3.52E+09	3.97E+09	2.61E-12	311.0	314.0	56.92	1.1794E-11	295.0	2.3393E-13	2.0683E-06	0.23	2.3811E-03
1960	2.04E+009	3.82E+09	5.60E+09	2.63E-12	313.4	316.9	46.98	1.2713E-11	296.0	2.3410E-13	2.0715E-06	0.32	2.3844E-03
1962	2.57E+09	4.55E+09	5.30E+09	2.66E-12	316.7	318.5	40.54	1.3866E-11	296.0	2.3430E-13	2.0738E-06	0.41	2.3878E-03
1964	2.69E+09	4.83E+09	6.15E+09	2.67E-12	318.0	319.6	46.22	1.4391E-11	296.0	2.3439E-13	2.0753E-06	0.42	2.3892E-03
1966	3.00E+09	4.71E+09	7.35E+09	2.70E-12	319.4	321.4	43.97	1.4959E-11	298.0	2.3448E-13	2.0769E-06	0.44	2.3906E-03
1968	3.29E+09	5.10E+09	7.98E+09	2.72E-12	321.2	323.0	39.08	1.5587E-11	298.0	2.3468E-13	2.0785E-06	0.46	2.3922E-03
1970	3.57E+09	5.51E+09	9.35E+09	2.73E-12	323.0	325.7	38.97	1.6273E-11	298.5	2.3479E-13	2.0801E-06	0.47	2.3937E-03
1972	4.05E+09	5.89E+09	1.02E+10	2.75E-12	325.3	327.4	37.10	1.7035E-11	301.0	2.3491E-13	2.0819E-06	0.49	2.3953E-03
1974	4.38E+09	6.62E+09	1.24E+10	2.78E-12	327.7	330.2	36.72	1.7878E-11	302.0	2.3502E-13	2.0837E-06	0.51	2.3969E-03
1976	4.62E+09	7.27E+09	1.11E+10	2.80E-12	330.3	332.0	34.65	1.8778E-11	303.0	2.3515E-13	2.0857E-06	0.53	2.4006E-03
1978	4.86E+09	7.87E+09	1.16E+10	2.83E-12	333.1	335.0	34.81	1.9726E-11	303.2	2.3528E-13	2.0899E-06	0.55	2.4026E-03
1980	5.19E+09	8.48E+09	1.22E+10	2.85E-12	336.0	338.7	37.30	2.0731E-11	303.0	2.3543E-13	2.0922E-06	0.58	2.4046E-03
1982	5.32E+09	9.05E+09	1.09E+10	2.87E-12	339.0	341.4	41.95	2.1782E-11	303.0	2.3558E-13	2.0945E-06	0.60	2.4067E-03
1984	5.11E+09	9.65E+09	1.09E+10	2.89E-12	341.5	344.6	43.77	2.2825E-11	303.0	2.3575E-13	2.0970E-06	0.62	2.4087E-03
1986	5.28E+09	9.92E+09	1.15E+10	2.92E-12	344.6	347.4	44.00	2.4953E-11	304.7	2.3594E-13	2.0996E-06	0.64	2.4109E-03
1988	5.61E+09	1.00E+10	1.22E+10	2.94E-12	346.9	351.6	44.11	2.6111E-11	307.3	2.3613E-13	2.1023E-06	0.66	2.4129E-03
1990	5.97E+09	1.04E+10	1.25E+10	2.97E-12	349.8	354.4	43.97	2.7323E-11	308.7	2.3633E-13	2.1051E-06	0.68	2.4148E-03
1992	6.15E+09	1.11E+10	1.26E+10	2.99E-12	352.8	356.4	44.19	2.8596E-11	308.5	2.3653E-13	2.1081E-06	0.70	2.4172E-03
1994	6.18E+09	1.11E+10	1.27E+10	3.02E-12	355.8	358.8	44.89	2.9903E-11	311.3	2.3673E-13	2.1111E-06	0.72	2.4195E-03
1996	6.29E+09	1.11E+10	1.29E+10	3.04E-12	358.8	362.6	46.09	3.1087E-11	313.7	2.3696E-13	2.1143E-06	0.74	2.4221E-03
1998	6.55E+09	1.13E+10	1.32E+10	3.07E-12	361.9	366.6	45.74	3.2406E-11	315.0	2.3718E-13	2.1176E-06	0.76	2.4248E-03
2000	6.64E+09	1.17E+10	1.35E+10	3.10E-12	365.0	369.5	45.64	3.3745E-11	316.7	2.3740E-13	2.1211E-06	0.78	2.4278E-03
2002	6.75E+09	1.21E+10	1.39E+10	3.12E-12	368.2	373.2	45.76	3.5118E-11	318.5	2.3764E-13	2.1246E-06	0.79	2.4308E-03
2004	6.98E+09	1.25E+10	1.65E+10	3.16E-12	371.6	377.5	42.29	3.6594E-11	320.8	2.3788E-13	2.1283E-06	0.81	2.4340E-03
2006	7.78E+09	1.30E+10	1.81E+10	3.19E-12	375.5	381.9	40.93	3.8207E-11	323.0	2.3813E-13	2.1322E-06	0.82	2.4373E-03
2008	8.35E+09	1.35E+10	1.91E+10	3.23E-12	379.8	385.6	40.44	3.9917E-11	324.6	2.3839E-13	2.1363E-06	0.84	2.4409E-03
2010	8.75E+09	1.42E+10	1.96E+10	3.27E-12	384.3	389.8	40.82	4.1693E-11	333.5	2.3866E-13	2.1405E-06	0.86	2.4445E-03
2012	9.01E+09		1.99E+10	3.31E-12	389.0		41.61	4.3522E-11		2.3894E-13	2.1449E-06	0.90	2.4485E-03

n	E(n)	T(n)	A(n)		p(n)				p*(n)		DIC	DTeq	ALK

Table A2.3 Model Predictions
700m sea depth with CO2 neutralization Model

year	c emission/yr c ml b sea	c ml b sea	co2 yr in air	Sum co2 in air	calc ppm	Obs ppm	% lo sea	total c used	p' sea	TotC in sea	D E	glob warm	Alk TA
A	B	C	D	E	F	G	H	I	J	K	L	M	Q
2010	9.01E+09	1.35E-10	1.96E-10	3.27E-12	389.0	389.8	40.82	4.1693E+11	324.6	2.3866E-13	2.1405E-06	0.88	2.4446E-03
2012	9.28E+09	1.42E-10	1.59E-10	3.31E-12	393.7		41.61	4.3522E+11	333.5	2.3894E-13	2.1449E-06	0.90	2.4485E-03
2014	9.55E+09	1.32E-10	2.18E-10	3.35E-12	398.9		37.78	3.8338E+11	334.0	2.3921E-13	2.1496E-06	0.92	2.4531E-03
2024	1.10E+10	1.43E-10	2.60E-10	3.61E-12	429.9		35.41	4.8906E+11	344.0	2.4064E-13	2.1738E-06	1.03	2.4766E-03
2034	1.26E+10	1.89E-10	2.75E-10	3.89E-12	459.1		40.74	6.0424E+11	360.0	2.4252E-13	2.2043E-06	1.14	2.5043E-03
2044	1.45E+10	2.26E-10	3.08E-10	4.09E-12	499.1		42.31	7.4016E+11	378.0	2.4478E-13	2.2414E-06	1.25	2.5391E-03
2054	1.67E+10	2.66E-10	3.47E-10	4.54E-12	540.4		43.46	8.9647E+11	402.0	2.4744E-13	2.2861E-06	1.36	2.5816E-03
2064	1.92E+10	3.04E-10	4.01E-10	4.94E-12	588.0		43.17	1.0762E+12	427.0	2.5049E-13	2.3393E-06	1.48	2.6345E-03
2074	1.73E+10	3.54E-10	2.80E-10	5.22E-12	621.4		55.82	1.2589E+12	447.0	2.5403E-13	2.4023E-06	1.55	2.6983E-03
2084	1.56E+10	3.84E-10	1.88E-10	5.41E-12	643.7		57.17	1.4233E+12	469.0	2.5787E-13	2.4722E-06	1.59	2.7708E-03
2094	1.40E+10	3.84E-10	1.30E-10	5.54E-12	659.1		74.76	1.5713E+12	483.0	2.6171E-13	2.5469E-06	1.62	2.8528E-03
2104	1.26E+10	3.87E-10	7.52E-09	5.63E-12	668.1		83.75	1.7044E+12	498.0	2.6559E-13	2.6249E-06	1.64	2.9409E-03
2114	1.14E+10	3.74E-10	4.22E-09	5.66E-12	673.1		89.85	1.8243E+12	489.0	2.6933E-13	2.7051E-06	1.65	3.0355E-03
2124	1.02E+10	4.05E-10	-3.02E-09	5.63E-12	669.5		108.07	1.9322E+12	479.0	2.7338E-13	2.7860E-06	1.64	3.1263E-03
2134	9.20E+09	4.19E-10	-8.18E-09	5.55E-12	659.8		124.26	2.0292E+12	483.0	2.7757E-13	2.8659E-06	1.62	3.2126E-03
2144	8.28E+09	3.89E-10	-8.53E-09	5.46E-12	649.6		128.11	2.1166E+12	500.0	2.8146E-13	2.9441E-06	1.61	3.3007E-03
2154	7.45E+09	3.29E-10	-5.60E-09	5.41E-12	643.0		120.48	2.1953E+12	513.0	2.8475E-13	3.0207E-06	1.59	3.3862E-03
2164	6.70E+009	2.86E-10	-4.00E-09	5.37E-12	638.2		116.28	2.2660E+12	506.0	2.8761E-13	3.0564E-06	1.58	3.4673E-03
n	E(n)	T(n)	A(n)	p(n)					p*(n)		DIC	DTeq	ALK

The model coding, in .XML format, can be obtained from the author chemeng@btinternet.com, or the website btinternet.com/~chemeng

References

Archer, D., Kheshgi H., Maier–Reimer E., *Multiple timescales for neutralization of Fossil Fuel CO_2,* Geo Res Letters, 24, 405–408, 1997

Archer D., Eby M., Brovkin VC., Ridgewell A., Cao L., Mikolajewicz U., Caldeira K., Matsumoto K., Munhoven G., Montenegro A., Tokos K., *Atmospheric Lifetime of Fossil fuel Carbon Dioxide,* Annu Rev. Planet Sci 37, 117–134, 2009

Feely,R.A., Sabine,C. L., Lee, K., Berelson, W., Kleypas, J., Fabry, V. J., Millero, F. J., *Impact of Anthopological CO_2 on the $CaCO3$ system in the Oceans,* Science, 2004, 305 362 – 366

Sabine, C.L., Feely, R. A., Gruber, N., Key, Lee, K., R. M., Billister, J. L., Wanninkhof, R., Wong, C. S., Wallace, D. W. R., Tilbrook, B., Millero, F. J., Peng, T., Kozyr, A., Ono, T., Rios, A. F.,*The Oceanic Sink for Anthropogenic CO_2* Science Vol. 305, no. 5682 pp. 367–371,

,

16 July 2004
Sigman, D. M., and Boyle, E. A.,(2000), *Glacial/interglacial variations in atmospheric carbon dioxide,* Nature, vol 407, 859

Appendix 2   The CO2 Neutralization Reaction

Alphabetical Index

www.ingramcontent.com/pod-product-compliance
Lightning Source LLC
Chambersburg PA
CBHW081107170526
45165CB00008B/2363